PHYSIOLOGY AND PATHOPHYSIOLOGY OF TEMPERATURE REGULATION

PHYSIOLOGY AND PATHOPHYSIOLOGY OF TEMPERATURE REGULATION

Editor

CLARK M. BLATTEIS

*The University of Tennessee
Memphis, TN, USA*

World Scientific
Singapore • New Jersey • London • Hong Kong

Published by
World Scientific Publishing Co. Pte. Ltd.
P O Box 128, Farrer Road, Singapore 912805
USA office: Suite 1B, 1060 Main Street, River Edge, NJ 07661
UK office: 57 Shelton Street, Covent Garden, London WC2H 9HE

British Library Cataloguing-in-Publication Data
A catalogue record for this book is available from the British Library.

Coventry University

PHYSIOLOGY AND PATHOPHYSIOLOGY OF TEMPERATURE REGULATION

Copyright © 1998 by World Scientific Publishing Co. Pte. Ltd.

All rights reserved. This book, or parts thereof, may not be reproduced in any form or by any means, electronic or mechanical, including photocopying, recording or any information storage and retrieval system now known or to be invented, without written permission from the Publisher.

For photocopying of material in this volume, please pay a copying fee through the Copyright Clearance Center, Inc., 222 Rosewood Drive, Danvers, MA 01923, USA. In this case permission to photocopy is not required from the publisher.

ISBN 981-02-3172-5

Printed in Singapore by Eurasia Press Pte Ltd

PREFACE

Body temperature is one of the clinical vital signs, along with pulse and respiratory rates. It is included in every initial medical history, it is monitored at regular intervals in all hospitalized patients, and, under certain conditions, it is also recorded continuously. This is so because a deviation of body temperature, up or down, is symptomatic of a pathologic condition. For example, fever is generally regarded as a hallmark of infection.

One would expect, therefore, that a subject so fundamental and relevant as the mechanisms of the control of body temperature would be covered in adequate and up-to-date detail in all textbooks of medical physiology. Unfortunately, that has not been the case for the past 15 years. With a very few and notable exceptions, the presentation of this important area of physiology has been minimal, outdated, and even, in too many instances, inaccurate, not only in textbooks published in English in the U. S. A. but also in other languages in different countries of the world. This state of affairs has been cause for increasing concern by the Commission on Thermal Physiology of the International Union of Physiological Sciences (CTP/IUPS), which is charged with surveying and fostering this particular field of physiology in the global setting; the membership of the CTP/IUPS consists of recognized experts from around the world in each of the subfields of thermal physiology who have contributed significantly to the knowledge of their specialty through research and have taught their specific as well as other areas of physiology for 20 years or more. In consequence, the Commission resolved in 1993 that publication of a relatively short, didactic monograph on thermal physiology was warranted to remedy the deficient coverage of the field. All the members agreed to contribute from the perspective of their expertise to produce a monograph readable and understandable by students. The objective was to develop a concise, yet not oversimplified text, free of advanced details, but sufficiently complete to provide pertinent, state-of-the-art information. Thus, this text is not intended to be all-encompassing, but rather to provide a concise, up-to-date, and relevant background to understanding how body temperature is controlled under the multivariate conditions normally encountered during life. It should be mentioned that one presumption was made by the contributing authors when writing their chapters, *viz.*, that the reader already possesses a suitably adequate background in mammalian physiology. That was done because thermoregulation does not operate through a well-defined and separate system of its own, but rather employs all of the systems of the body and integrates their activities into appropriate and coordinated reactions such that a stable temperature is maintained even under conditions that, *a priori*, might be expected to cause body temperature to become deviated. Preliminary familiarity with the physiology of these systems is, therefore, necessary.

Planning, organization, and initial outlining of individual chapters of the book took place during 1994. The working draft was completed late in 1995. It took another year and a half to write the final, complete text; each chapter essentially underwent two revisions before its definitive version. The book is offered now in the hope that it will meet the requirements of medical and other health sciences students preparing for their careers. For indeed, although there is no "thermoregulation disease" *per se* from which one may suffer chronically, there are nevertheless many conditions that secondarily result in permanently high or low body temperature, as, for example, certain brain lesions or tumors, various endocrinopathies, or addiction to drugs of abuse. Conditions also exist that can overwhelm the ability of the body to maintain its temperature within the normal ranges, such as exposure to extreme heat or cold. These conditions can result in fatal outcomes if intervention is not prompt and appropriate. Every physician is likely to come into contact at some time with patients experiencing one or another such acute thermoregulatory deficiency. The correct understanding of the mechanisms controlling body temperature should facilitate detection and treatment of conditions in which body temperature is disordered.

To prepare students for this eventuality, this book is divided into two parts. Part 1 deals with the physiological principles that underlie the maintenance of body temperature. Basic topics covered include the biophysics of heat transfer, the mechanisms of heat production and loss, the neural control of body temperature, behavioral temperature regulation, temperature regulation during exercise, and the control of body temperature in relation to age; the latter three topics are particularly relevant to practical settings and are usually not addressed in other texts. Part 2 provides current information on the pathophysiological bases and clinical manifestations of disordered temperature regulation, including fever and the thermal effects of exposure to environmental extremes. The thermoregulatory effects of certain diseases and various drugs are also considered. Practitioners too may thus find this monograph useful as a refresher to update their knowledge and to devise proper and rational patient care. The book contains 13 chapters, each contributed by the expert on the topic being developed. Each chapter is divided into sections describing the relevant principles. To assist the reader in mastering its contents, a few pedagogic features are included in each chapter, such as a set of learning objectives, "bullets" that emphasize the important components of the concepts being presented, key words, a list of suggested reading in addition to the pertinent references, a self-test, and, of course, appropriate tables and illustrations. The editor is deeply indebted to his colleagues for their hard work and generous dedication to this project.

Readers may note that some chapters are written in a slightly less idiomatic English than others. This is due to the diverse national backgrounds of their authors. Thus, the wording of certain phrases, the spelling of certain terms, or the syntactic style and punctuations are not uniform throughout the book. The editor has deliberately refrained from imposing his own, personal style on his co-authors. His only concern was that the concepts being presented be expressed in the clearest manner possible, in the words chosen by each contributor. Every author took care to insure that the informational content of his or her chapter was as accurate and up-to-date as practicable. Some redundancies do occur among certain chapters, but they are intentional so that the reader may recall and then put the relevant concepts into each new context. The editor hopes that he has not overlooked any major gaps, although in a monograph of this type, complete, detailed coverage of all topics is not possible or, for that matter, appropriate or necessary. Among the topics, hence, not covered in great detail are the roles of peripheral and central neurochemical mediators in temperature regulation, a still very confused subject in any case. However, the suggested reading list that follows the references in each chapter should be a guide to such issues. Specific questions can also be addressed by interested readers to thermal physiologists around the world via the Thermophysiology World-Wide Web Server at http://physio1.utmem.edu/THERMOPHYSIOLOGY/.

Clark M. Blatteis, Ph.D.
University of Tennessee, Memphis
September 1997

LIST OF CONTRIBUTORS

Blatteis, Clark M.
Department of Physiology and Biophysics
The University of Tennessee
Memphis 38163
USA

Boulant, Jack A.
Department of Physiology
Ohio State University
Columbus, Ohio
USA

Cabanac, Michel
Départment de Physiologie
Faculté de Médecine
Université Laval
Québec, G1K 7P4
Canada

Cannon, Barbara
University of Stockholm
Sweden

Cooper, Keith E.
Department of Medical Physiology
Faculty of Medicine
University of Calgary
3330 Hospital Drive N. W.
Calgary, Alberta
Canada T2N-4N1

Hales, J. R. S.
The University of Sydney
Faculty of Veterinary Science
Camden, N. S. W.
Australia

Horowitz, Michal
The Hebrew University
Hadassah Medical School
Department of Physiology
91120 Jerusalem
Israel

Jan Nedergaard
University of Stockholm
Sweden

Jansky, Ladislav
Faculty of Science
Charles University
Prague
Czech Republic

Kaciuba-Uscilko, Hanna
Department of Applied Physiology
MRC, Polish Academy of Sciences
Warsaw
Poland

Laburn, Helen P.
Department of Physiology and Brain Function Research Unit
University of the Witwatersrand
Johannesburg
South Africa

Mercer, James B.
Department of Medical Physiology
Institute of Medical Biology
University of Tromsoe
N-9006 Tromsoe
Norway

Milton, A. S.
University of Cambridge
England

Morimoto, Taketoshi
Department of Physiology
Kyoto Prefectural University of Medicine
Kamigyoku
Kyoto 602
Japan

Nielsen, Bodil
August Krogh Institutet
University of Copenhagen
Denmark

Werner, Jürgen
Ruhr-University
Medical Faculty
Center for Biomedical Methods
MA 4/59, D-44780 Bochum
Germany

Zeisberger, Eugen
Physiologisches Institut
Klinikum der Justus Liebig Universität
Aulweg 129, D-35392 Giessen
Germany

CONTENTS

Preface		v
List of Contributors		ix

Part 1 Physiology — 1

Chapter 1 Thermal Physiology: Brief History and Perspectives (*Milton*) — 3
2 Body Temperature (*Blatteis*) — 13
3 Biophysics of Heat Exchange between Body and Environment (*Werner*) — 23
4 Heat Production Mechanisms
   4.1. Shivering (*Jansky*) — 47
   4.2. Nonshivering Thermogenesis and Brown Adipose Tissue (*Cannon and Nedergaard*) — 61
5 Heat Loss Mechanisms (*Morimoto*) — 79
6 Neural Thermal Reception and Regulation of Body Temperature (*Boulant*) — 93
7 Thermiatrics and Behavior (*Cabanac*) — 107
8 Temperature Regulation in Exercise (*Nielsen and Kaciuba-Uscilko*) — 127
9 Body Temperature and Age
   9.1. Thermoregulation in the Neonate (*Laburn*) — 145
   9.2. Thermoregulation in the Elderly (*Cooper*) — 161

Part 2 Pathophysiology — 175

Chapter 10 Fever (*Blatteis*) — 177
11 Thermoregulatory Consequences of Prolonged Exposure to Thermal Extremes
   11.1. Prolonged Exposure to Heat (*Horowitz*) — 193
   11.2. Cold Adaptation (*Zeisberger*) — 207
12 Pathophysiological Consequences of Exposure to Thermal Extremes
   12.1. Pathophysiology of Hyperthermia (*Horowitz and Hales*) — 229
   12.2. Hypothermia and Cold Injuries in Man (*Mercer*) — 245
13 Temperature Regulation in Special Situations (*Blatteis*) — 259

Appendix 1	Chapter 2 Methods of Body Temperature Measurement (*Blatteis*)	273
Appendix 2	Chapter 3 Table B: List of Symbols (*Werner*)	281
Index		285

PART 1
PHYSIOLOGY

Chapter 1

THERMAL PHYSIOLOGY: BRIEF HISTORY AND PERSPECTIVES

A. S. MILTON
University of Cambridge, England

"If a definition of life were required, it must be most clearly established on that capacity, by which the animal preserves its proper heat under the various degrees of temperature of the medium in which it lives. The most perfect animals possess this power in a superior degree, and to the exercise of their vital functions this is necessary. The inferior animals have it in a lower degree, in a degree however suited to their functions. In vegetables it seems to exist, but in a degree still lower, according to their more limited powers, and humbler destination.....

There is reason to believe, that while the actual temperature of the human body remains unchanged, its health is not permanently interrupted by the variation in the temperature of the medium that surrounds it; but a few degrees of increase or diminution of the heat of the system, produces disease and death. A knowledge therefore of the laws which regulate the vital heat, seems to be the most important branch of physiology".

James Curry 1808

1. Introduction

How true are those words of James Currie spoken nearly 200 years ago! Modern civilization as we know it today would not exist if man had not learnt to maintain his body temperature in adverse climatic conditions by inventing and wearing clothes. In addition, man learnt to protect himself from the elements by making shelters, and learnt one other skill which separates man from all other animals, the use of fire to provide warmth. This allowed him to move out of the tropics, where he had originated, into colder climates. Let us not forget that man is a tropical animal. A naked man can survive in a warm climate. If the temperature gets too hot, then vasodilatation and sweating occur, and man may move to a cooler place, an example of behavioural thermoregulation. However if the ambient temperature begins to fall, man runs into difficulty: a few degrees fall and vasoconstriction and shivering will occur, a few further degrees lower and a naked man will die. Indeed, without clothing, mankind would still be composed of naked savages in tropical Africa or perhaps Australasia.

Whereas in lower forms of life survival is possible over quite wide variations in deep body temperature, in mammals the limits are much narrower. Lower animals do alter their body temperature, and there are optimal temperatures for them to function properly. An interesting example occurs in some moths who cannot fly until their body temperatures are over 30°C. They can raise their temperatures by vibrating their wings; this muscular energy produces heat. If, however, the ambient temperature falls so low such that the moth loses heat so rapidly that it cannot raise its body temperature enough, then it is unable fly and therefore cannot mate, and the species is endangered. As well as changing deep body temperature by producing energy, so called 'cold blooded' animals can change their temperature by behavioural means, such as moving into hotter or colder environments. During any one day, the deep body temperature may vary over several degrees. However, even lower forms of animals appear to have a 'preferred' ambient temperature.

2. Homeothermy and Poikilothermy

John Hunter (1837) was the first person to point out that animals should properly be divided into those with constant temperatures and those with variable temperatures, and not into warm blooded and cold blooded; and it was Bergmann who, in 1847, introduced the terms homeothermic and poikilothermic, respectively, to describe the two divisions.

The actual body temperature of any individual is determined by two parameters, namely, heat gain and heat loss. Heat gain is primarily linked to heat production, in the form of metabolic energy. Heat production in man is fairly constant; i.e. related to the basal metabolic rate and may be increased during exercise and during the process of shivering when uncoordinated muscular movements produce heat. Heat loss is related to radiation, convection and conduction from the external surfaces, and may be varied by changes in blood flow to the skin. These three are obviously considerably affected by clothing in man, and by the presence of fur or feathers in other animals and birds, respectively. Mammals also make use of evaporative heat loss from sweating, spreading of saliva and panting. Evaporative heat loss is the only way an animal can lose heat when ambient temperature is above body temperature. In man, sweating is of particular importance, whereas in animals with thick fur, panting is more common.

In 1780, Lavoisier demonstrated that the heat produced by the body could be explained by what he referred to as "combustion processes" which culminated in the production of carbon dioxide. Lavoisier believed that the heat was produced in the lungs.

The eccrine sweat glands of man were first observed by Malpighi in 1687. In 1758, Ellis noted that the human skin had a temperature of 97°F when the ambient temperature was 105°F. The importance of this observation was provided by Franklin who suggested that the lower skin temperature was due to the evaporation of sweat. It was Langley who, in 1891, mapped out the sympathetic nerve supply to the sweat glands. In man, sweating is the most important method of heat loss; indeed, man occupies the high point in the development of sweating in the animal kingdom. Of particular interest is the fact that in man sweating from the eccrine glands is under sympathetic cholinergic control. This is in contrast to most of the rest of the animal kingdom where the sweat glands are under adrenergic control. Lovat Evans showed, for example, that in the horse, the apocrine glands, which are sweat glands in this species, are stimulated by circulating adrenaline from the adrenal medulla and inhibited by sympathetic nerve stimulation.

The role of the hypothalamus as the centre for temperature control was delineated over the first part of this century. The experiments of Brodie (1811) first demonstrated that the central nervous system affected body temperature. The site was localized to the hypothalamus, with much of the evidence coming from lesion studies in a variety of animals including man. Of particular importance are the works of Barbour, of Cannon and of Ranson. The concept thus gradually developed over the years that the various changes which occurred to maintain a relatively constant temperature, such as sweating, panting and changes in vasomotor tone, which are of course autonomic responses, are controlled by the hypothalamus. Shivering, which is a somato-motor response is also controlled by the hypothalamus.

It was Claude Bernard who first proposed 'le milieu intérieur 'in 'Leçons sur la chaleur animale' to describe the internal regulatory mechanism of the body. Bernard was a pupil of Magendie. In 1844, Magendie and Bernard collaborated in an enquiry into what they termed 'animal heat'. This enquiry dated back to the work on Gustav Magnus who, in 1842, had published with Lavoisier that the oxidation of respiration took place in the lungs. However, Bernard showed that temperature probes inserted into the heart of the horse recorded the temperature of the blood in the right ventricles as warmer than that in the left ventricle, showing that the blood going to the lungs was warmer than that coming from them, nullifying Lavoisier's statement. Bernard and others also showed that, in several internal organs of the body, the venous blood was warmer than the arterial, indicating that these tissues were the site of oxidation and, thus, heat production.

In his concept of le milieu intérieur, Claude Bernard placed the emphasis upon the protection furnished living cells by the nearly constant composition and temperature of the fluids which bathe them.
"All the vital mechanisms, varied as they are, have only one subject, that of preserving constant the conditions of life in the internal environment. The fixity of the milieu supposes a perfection of the organism such that the external variations are at each instant compensated for and equilibrated. Therefore, far from being indifferent to the external world, the higher animal is on the contrary constrained in a close and masterful relation with it, of such fashion that its equilibrium results from a continuous and delicate compensation established as if by the most sensitive of balances. All of the vital mechanisms however varied they may be, have always but one goal, to maintain the uniformity of the conditions of life in the internal environment." (translated from the French)

In the very simplified model of temperature regulation described here, of particular importance is what determines the actual body temperature and how is it controlled in man at around 37.5 °C. As in most physiological processes, a stimulus is received and a response is initiated. In temperature regulation, thermosensitive neurones that have thermosensitive receptors (nerve endings) and which are present in the periphery, in deep organs of the body and in the central nervous system, respond to changes in temperature. So called 'cold sensitive' neurones increase their firing rates as temperature falls, and 'heat sensitive' neurones increase their firing rates as temperature rises. The afferent input from these neurones is received and processed in the brain in the 'so called' thermoregulatory centre thought to be located in the preoptic area of the anterior hypothalamus (PO/AH), where the appropriate efferent responses controlling sweating, shivering and changes in vasomotor tone are activated to maintain normal body temperature. This is in no way dissimilar to the control of blood pressure where afferent signals from baroreceptors activate the appropriate responses to the cardiovascular system to maintain a normal blood pressure. The outcome, then, of the afferent input and efferent response is to maintain a fairly constant temperature, though coupled with a circadian rhythm. In man, a maximum temperature is seen in mid-afternoon and a minimum in the early hours of the morning, during sleep. Many physiologists talk about a 'set point' in the hypothalamus which is responsible for maintaining deep body temperature. They suggest that any deviation from the set point results in thermoregulatory responses to restore the temperature to the set point temperature. This is an engineer's concept, which should be suspect to the physiologist, as would be the concept of a 'barostat' in the brain determining blood pressure.

3. Thermometry and Fever

Until the development of the clinical thermometer, deep body temperature could not be measured. Assessment of body temperature was subjective, based on the feel of the skin, on whether it felt hot or cold, and on the sensations of the person concerned, whether he or she felt hot or cold. Even under these conditions, one abnormality in body temperature was recognized and has been since the beginning of the human race, and that is fever. Illness and fever have always been associated; even today the first action of any physician on visiting a patient is to insert a thermometer into the patient's mouth. This, of course, fulfils two objectives ; firstly to see whether the patient has a fever and therefore is probably ill, but secondly and perhaps more importantly to render the patient speechless while the doctor makes his initial examination!

The symptoms of fever - shivering, cold and clammy extremities, the burning forehead, profuse sweating and the subjective feeling of heat and cold - have been recorded throughout history. The very words we use to describe fever stem from classical languages: fever from the Latin *febris*, and pyrexia and pyretic from the Greek πψρετοσ. The folklore of fever is immense, with every culture and civilization having its own myths and explanations. The best known of the early explanations are recorded in the works of Empedocles and Hippocrates. Empedocles of Agrigentum ,in Sicily (477 - 432 B.C.), proposed the doctrine of the four elements, namely earth, air, fire and water, as the "four-fold root of all things." The human body was considered to be composed of these four fundamental elements, with health resulting from a correct balance of all four, and disease an imbalance. Plato and Aristotle introduced the idea of four qualities, 'dry, cold, hot and moist", and combined them with the four elements into the following scheme: cold and dry represented earth, hot and moist represented air, hot and dry represented fire, and cold and moist represented water. Hippocrates put forward the idea of the four humours: blood, phlegm, yellow bile and black bile; in which the scheme of the four elements was modified such that cold and dry represented black bile, hot and moist represented blood, hot and dry represented yellow bile, and cold and moist represented phlegm. Hippocrates maintained that illness resulted from an overproduction of one of these four humours and that the body destroyed the excess humours, perhaps by increasing the body's heat. (i.e. by developing a fever). As Currie says of Hippocrates: "Perceiving the increase of heat to be the most remarkable symptom in fever, he assumed this for the cause and founded his distinctions of fevers on the different degrees of the intenseness of this heat. He had no instruments that could measure this exactly, and necessarily trusted to his sensations." Hippocrates obviously believed that fever was necessary for the reduction of disease. In his book *Methodus medendi*, Galen (131-201 A.D.) stated that medicines should be classified according to their various humoral content. Perhaps because of the teachings of Hippocrates and Galen, which considered fever to be 'beneficial', the use of drugs as antipyretics does not appear in their writings, or indeed in the writings of any physician up until the eighteenth century. Any effect of a drug on body temperature was purely accidental to its curing of the disease. In the seventeenth century one of the most eminent medical man of his day, Thomas Sydenham (1624 - 1689), revised the Hippocratic methods of observation and stated "Nature calls in fever as her usual instrument for expelling from the blood any hostile matters that may lurk in it". Quoting once again from Currie, "It was the postulate of Sydenham that every disease is nothing else but an endeavour of nature to expel morbific matter of one kind or another, by which her healthy operations are impeded". However, from the late nineteenth century right up until to today, physicians have generally regarded fever as being harmful to the body and have made every effort to reduce it. Perhaps as we approach the millennium, we should revise our ideas ; we now know that fever is just one aspect of the acute phase of the immune response, so unless the increase in body temperature becomes life threatening, it may be best to let the fever run its course. Since, therefore, fever is part of the body's response to infection and tissue damage, should we not treat the cause of the fever and not the fever itself?

Modern studies on thermophysiology and our understanding of fever stem from two important developments. The first was the introduction of thermometry and the clinical thermometer, and the second was the introduction of antipyretic drugs. The first thermometer is properly attributed to Galileo, who is said to have invented it sometime between the years 1593 and 1597. Galileo's thermometer consisted of a glass bulb connected to a tube; the bulb was inverted and the end of the tube placed in a bowl of wine, such that the wine came part way up the tube. When the bulb was warmed, the expansion of the air forced the level of the wine in the tube to fall and conversely when the bulb was cooled, and as the air contracted, the wine rose in the tube. The first clinical thermometer was described by Sanctorius; it was similar to the one invented by Galileo, but Sanctorius used it to measure changes in temperature in patients. The warmer the patient, the faster the wine level fell in the tube. It was not until the second decade of the eighteenth century that the measuring thermometer was described by Fahrenheit. The importance of thermometry was described by Wunderlich (1815 - 1877). In a book on his work, Seguin states: "Thernometry has truly discovered, according to the vivid expression of Wunderlich, a new world, the one dreamed of by Currie, the law of the action of external upon human temperature. But this therapeutic application of the two relative terms of caloric to the treatment of disease is only the initial impulse of an immense revolution, whose subsequences, hidden to the view of the far seeing Currie, are hardly traceable in our horizon; I mean. the calorific and frigorific action of all our medicines, vegetables, and their alkaloids, metals, metalloid bodies, and gases. This entirely new field of observation, and of therapeutic action, could vanish like a mirage if thermometry could be suppressed. But, far from this impious impossibility, thermometry will find out even the positivism of empiricism in the law of concordance of the apparently most discordant treatments; and will reconcile schools which were divided, only because they did not know that their far diverging means converged to the same action and object - the keeping up of normal temperature, that is to say, life; and the suppressing of the sources of pathological temperatures - that is, death, *in propria persona.* "

The second important development was the use of drugs as antipyretics. The Reverend Edward Stone of Chipping Norton in Oxfordshire, England published a paper in 1763 in the Philosophical Transactions of the Royal Society, which has the title 'An account of the success of the bark of the willow in the cure of agues'. This was the first scientific paper to be published which describes the use of an antipyretic drug. However, there are records of the use of decoctions of willow bark as a 'general febrifuge' which go back many centuries. The use of such concoctions continued throughout the medieval period until the introduction of synthetic salicylates at the end of the nineteenth century. However, it would appear that whereas the medical profession brought up in the teachings of Hippocrates and Galen did not use drugs as 'febrifuges' and did not wish to use medicines which brought down fever, their use as such was part of folk medicine and used by healers outside of the medical profession.

There is an account of the use of willow bark as a successful substitute for cinchona bark in the Napoleonic Wars (1803 - 1815) in the treatment of fever; and, in 1798, Longmore wrote a paper on 14 soldiers who were poisoned in Quebec after consuming a "decoction of certain plants" in which he records that an extract of gaultheria (Wintergreen which contains methyl salicylate) "is frequently used by Canadians and is said to be cooling and grateful ptisan in fevers." Until 1874, quinine, derived from cinchona bark (also known as Peruvian bark), was the main antipyretic agent, its use of course, being primarily in the treatment of malarial fever.

It is interesting to remember that extracts of plants such as willow, which we now refer to as salicylates , had been used for centuries in medicine, but that no mention of their antipyretic actions were recorded until the paper of Stone in 1763. For example, Hippocrates recommended juice of the poplar for eye diseases and the leaves of willow trees in childbirth. Celsus, in the first century A.D., used juices of willow trees for removing corns, and Discorides recommended their use for earache, skin disease and gout. Galen suggested their use in the treatment of bloody wounds and ulcers. Similarly, one may find mention of many plants containing salicylates in the herbals of the Middle Ages and Renaissance. Neither Currie nor Wunderlich mention antipyretic drugs in their writings; the former regarded direct cooling of the body as the method of treatment of fever, and Wunderlich was primarily concerned with observations made during various fevers and the progress of the disease.

4. Conclusion

Just as this chapter began with the words of James Currie, so it is fitting to end with the words that he wrote in the Preface to his work and which he addressed to the Right Honourable Sir Joseph Banks:

"About eighteen years ago when I was in Edinburgh I discovered that the accounts given of the temperature of the human body under disease, even by the most approved authors, are, with a few exceptions, founded, not on any exact measurements of heat, but on the sensations of the patient himself, or his attendants.
Impressed with the belief, that till more accurate information should be obtained respecting the actual temperature in different circumstances of health, and disease, no permanent theory of vital motion could be established, nor any certain progress made in the treatment of those diseases in which temperature is diminished or increased."

Your faithful and very obedient Servant

James Currie

Liverpool, 31st October 1797

Suggested Reading

Bernard, C. (1859) Leçons sur les propriétès physiologiques et les altérations pathologiques des liquides de l'organisme. J.B. Ballière. Paris

Currie, J. (1808). Medical Reports on the effects of water, cold and warm, as a remedy in fever and other diseases, whether applied to the surface of the body, or used internally. 4th. London edition Philadelphia: Printed for James Humphreys and for Benjamin and Thomas Kite.

Milton, A.S. (ed) (1982). Pyretics and Antipyretics. Handbook of Experimental Pharmacology. vol. 60. Springer -Verlag; Berlin, Heidelberg.

Wunderlich, C.A. & Seguin, E. (1871). Medical thermometry and human temperature. William Wood, New York.

General References

Handbook of Physiology. Section 4. Adaptation to the Environment. eds: Dill, D.B., Adolph, E.F. and Wilber, C.G. American Physiological Society, Washington. D.C. 1964

Handbook of Physiology. Section 4, Environmental Physiology, vols 1 and 2. eds: Fregly M.J and Blatteis C.M. Oxford University Press New York 1996

Chapter 2

Learning Objectives
1. Explain the significance of body heat.
2. Know the factors determining the thermal state of the body.
3. Know the ranges, distribution, and variation of the "normal" body temperature.
4. Describe the terms and know the symbols used in the current nomenclature of thermal physiology and pathophysiology.
5. Define the terms and know the symbols used in the current nomenclature of thermal physiology and pathophysiology.
6. Describe the functional organization of the thermoregulatory system.

Bullets
B-1 Body heat is an end-product of energy transformation.
B-2 T_c is regulated.
B-3 A constant T_c imparts survival benefits.
B-4 The body has many different temperatures.
B-5 T_c is relatively stable.
B-6 The "normal" T_c is not precisely 37°C!
B-7 The T_c varies with the site measured.
B-8 Which temperature truly represents "body" temperature?

Chapter 2

BODY TEMPERATURE

CLARK M. BLATTEIS
Department of Physiology and Biophysics
The University of Tennessee, Memphis 38163, USA

1. Introduction

Internal body, or **core**, **temperature** (T_c) is one of the clinical vital signs, along with pulse and respiratory rates. It is included in every initial medical examination, it is followed at regular intervals in patients, and under certain circumstances it is also recorded continuously. This is so because a deviation of T_c from its normal limits, up or down, is pathognomonic, *i.e.*, symptomatic of a pathologic condition. For example, *fever* is generally regarded as a hallmark of infection.

Although there is no "thermoregulation disease," in the sense of "heart disease," from which one may suffer chronically, there are, nevertheless, many conditions that secondarily result in permanently high or low T_c, as for example, certain brain lesions or tumors (*e.g.*, in the hypothalamus or corpus callosum), various endocrinopathies (*e.g.*, polycythemia vera, hypo- or hyperthyroidism), addiction to drugs (*e.g.*, cocaine, alcohol). Conditions also exist that can overwhelm the ability of the body to maintain its temperature within the normal range, such as exposure to extreme heat (*heat illnesses, hyperthermia*) or cold (*cold injuries, hypothermia*). These conditions can be lethal if intervention is not prompt and appropriate. Every physician is likely to come into contact at some time with patients experiencing acute thermoregulatory deficiencies. Detection and treatment will be facilitated by an understanding of the mechanisms controlling T_c.

Thermoregulation does not operate through a well-defined and easily recognizable system of its own, such as do circulation, respiration, etc. Rather, it employs all of the systems of the body and integrates their activities into appropriate and coordinated reactions such that a stable body temperature is maintained under most conditions. The astute student will recognize, however, that, in fact, this situation is not unique, since the regulation of most physiological functions involves the participation of multiple systems and sub-systems, *e.g.*, salt-water homeostasis, blood glucose regulation. In this regard, thermoregulation is another excellent model in which the integrative physiology of the body is exemplified.

2. Significance of body temperature

The body has a "temperature" because combustion processes are constantly taking place in living organisms, releasing energy. A part of this energy is conserved as transformation, the rest is evolved as heat which, hence, accumulates within the body, causing its temperature to rise. Since heat, like fluids, flows down a gradient (Newton's law of cooling), the heat thus generated can readily be dissipated into the environment so long as it is cooler than the body. But what if the environment is warmer than the body? Should the heat then not accumulate? Moreover, since ambient temperature (T_a) varies widely, should T_c not be changing constantly as a function of T_a? Yet, clearly, the T_c in man and most mammals is relatively constant at *ca.* 37° C! This stability implies, therefore, that the heat produced in the body and that lost from it stay in relative balance, despite the large variations in T_a. This, in turn, suggests that both heat production and heat loss are regulated in some way. It may be further inferred that this control has important implications for survival. Indeed, since the rates of all chemical reactions, up and down, are largely dependent on temperature (van't Hoff-Arrhenius law), the ability to maintain T_c around a level presumptively such that biochemical reactions can operate optimally and bodily functions can proceed efficiently may be expected to free thus empowered animals from the vagaries of T_a; a freedom that would allow them to leave what would otherwise be a thermally restrictive environment better to explore, hunt, avoid predators, etc., *i.e.*, to survive.

B-2

Animals that possess the ability to regulate their T_c in this fashion are called **homeotherms**. Mammals and birds are homeotherms that regulate their T_c mainly by **autonomically** maintaining a high rate of basal heat production (*tachymetabolism*); they are called **endotherms**. Reptiles, on the other hand, are homeotherms that regulate their T_c mainly by **behaviorally** exchanging heat with their environment; they are called **ectotherms**. Organisms without effective temperature regulation by neither autonomic nor behavioral means are **poikilotherms** (*e.g.*, most insects); their T_c thus varies as a proportional function of T_a.

B-3

3. The "normal" body temperature

The heat content of the body can be measured in various ways: 1) as a quantity of heat, specifically as that needed to increase the temperature of 1 kg of water from 15 to 16 °C, *i.e.*, 1 kilocalorie or 1 kilojoule (kilocal x 4.187); 2) as a rate of heat exchange, *viz.*, per unit time and effective heat-exchanging area, *i.e.*, kilocal · h^{-1} · m^{-2} or Watts · m^{-2} (kilocal · h^{-1} · m^{-2} x 1.163); or 3) as a degree of sensible heat or cold, reckoned from melting ice to boiling water and expressed on a graded scale, *i.e.*, as temperature (in °C [preferred] or °F). Temperature, of course, is the easiest variable to measure and, consequently, it is the one most commonly monitored.

3.1 Distribution

The temperatures of different sites of the body vary significantly from site to site. Thus, temperatures taken at the surface are considerably different from those measured in an internal region of the body. They also vary depending on the specific body locus, superficial or deep. For example, in an individual at rest, the skin temperature of the trunk is normally higher than that of the arms, and the temperature of the heart is higher than that of the kidneys. However, the temperatures within the body are more uniform and stable over time than those at the surface, which are not uniform and can fluctuate widely and rapidly. In 1958, Aschoff and Wever [1] introduced the terms "thermal core," consisting of the intracranial, intrathoracic, and intraabdominal contents, the temperatures of which remain quite constant over a narrow range, and "thermal shell," including the skin and subcutaneous tissue and the limbs, the temperatures of which can change greatly.

The important derivation from this conceptual compartmentalization is as follows. As already pointed out, the energy expended within the core is not stored but rather dissipated into the environment, *ipso facto* primarily from the body's surface. It follows, therefore, that the temperature of the skin (T_{sk}) is a function of the quantity of the heat that flows into it. Since the rate at which heat is transferred from the surface to the environment is determined by the temperature difference between the two and the thermal conductivity of the medium (*e.g.*, air, water) through which the transfer occurs (see Chap. 3), it follows further that heat loss is promoted when the skin is warm and impeded when it is cool relative to the environment. One may infer, therefore, that, in order to maintain T_c stable, the rate of heat flow from core to

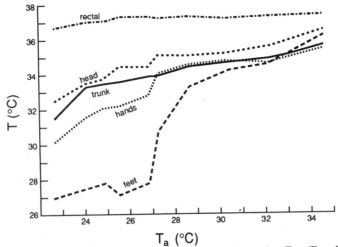

Fig. 1. T_c (rectal) and T_{sk} (various sites) of a subject exposed to various T_as. (From [4]).

shell is adjusted according to the body's thermal needs and that, as a result, T_{sk} varies more widely than T_c in relation to T_a (Fig. 1). Under steady-state conditions in a **thermoneutral** environment (*i.e.*, one in which neither the mechanism for heat production nor for heat loss is activated and the perceived thermal comfort is optimal), T_c thus is higher than T_{sk}. For resting, naked adults, this zone of T_a lies between 28 and 30 °C.

B-5

In addition to the thermal gradient, heat exchange is also importantly a function of the ratio of the effective heat-exchanging surface area to the volume (read: size or *mass*) of the exposed part. Thus, long, thin appendages with large surface area/volume ratios are better conduits for heat exchange than short, stubby bodies, *e.g.*, in humans, the arm relative to the trunk, the hands relative to the arms, and the fingers relative to the hands. This property accounts for the lower temperature of the extremities than of the trunk (Table 1).

Table 1. Relation between regional surface areas (A) and usual skin temperatures (T_{sk}) in adult man at rest in a thermoneutral environment with no wind and low humidity.

Region	A, m^2	% of total body A	T_{sk}, °C
Head	0.20	6	34.6
Trunk	0.70	36	34.6
Arms	0.10	8	33.0
Forearms	0.08	6	30.8
Hands	0.07	5	28.6
Thighs	0.33	19	33.0
Calves	0.20	13	30.8
Feet	0.12	7	28.6
TOTAL	1.80 (Total)		33.0 (Mean)

Indeed, some organs in various species are specialized on this basis for heat exchange, *e.g.*, the ears of rabbits, the tails of rats, the tongues of dogs. In humans, this feature also makes the distal limbs vulnerable in very cold environments (see Chap. 12.2).

3.2 Ranges and variation of T_c

The average T_c (measured orally [T_{or}]) of healthy (US) adult humans at rest in thermoneutrality is 36.8 ± 0.4 °C (98.2 ± 0.7 °F), with an average **nychthemeral** (*i.e.*, over an exact 24-h period, consisting of a day and a night) variability of 0.5 °C, being lowest at 6 AM and highest at 4-6 PM; in general, the T_{or} of women is slightly higher (36.9 °C) than that of men (36.7°C). The median for all subjects is 36.8 °C (98.2 °F), and the mode is 36.7 °C (98.0 °F). No statistically significant difference is demonstrable between Caucasian and Afro-American subjects [2].

B-6

Table 2. Some sources of variation of the "normal" core temperature.

EXOGENOUS	ENDOGENOUS
CLIMATOLOGICAL ENVIRONMENT (T_a; *barometric pressure; relative humidity*)	RHYTHMS (*nychthemeral; seasonal*)
PERIPHERAL INSULATION (*type and placement of clothing*)	GENDER (*menstrual cycle; pregnancy*)
DIET (*diet-induced thermogenesis; amount and composition of food consumed; time since last meal*)	AGE AND BODY SIZE
PHYSICAL ACTIVITY (*type and intensity; time since cessation of exercise*)	SUBCUTANEOUS INSULATION (*fat; blood flow*) WATER CONTENT
DRUGS	PSYCHOLOGICAL STATE

Under usual conditions, a variety of exogenous and endogenous factors affect T_c (Table 2); as these will be considered in detail in subsequent chapters, they will not be discussed here. Despite these factors, T_c does not fluctuate normally more than ± 2 °C around its 24-h mean. The condition in which T_c is within ± 1 standard deviation of this range is called **cenothermy** (also **normothermy**). Conditions in which T_c is below or above the estimated lower and upper limits of cenothermy are termed *hypothermia* and *hyperthermia*, respectively (see Chap. 12). In this regard, *fever* would be any T_c above the average maximum 37.2 °C (98.9 °F) recorded at 6 AM, or 37.7 °C (99.9 °F) recorded at 4 PM (Chap. 10).

4. Methods of body temperature measurement

In clinical practice, T_c is usually obtained by inserting a thermometric device into a natural orifice of the body, which should always be specified [3]. The choice of the orifice is mostly a matter of practicality and of esthetic preference. Typically, these include the mouth, the rectum, the ear canal, and, more invasively, the deep esophagus. The thermometric devices and techniques for obtaining accurate measurements, and the advantages and disadvantages of each site are described in Appendix 1. It should be appreciated from the preceding, however, that the value of T_c thus measured is a *representative*, not the true nor the mean T_c, albeit it is commonly used as a first order approximation of the mean body temperature.

B-7

5. Organization for thermal homeostasis

The ability of homeotherms to maintain their T_c within an optimal range despite large variations in T_a implies that the wherewithal for sensing temperature and translating this information into signals to effectors of heat production and/or heat loss functions exist in the body. This is, in fact, the manner by which the regulation of T_c is achieved; *viz.*, **thermosensors** continuously monitor the state of the controlled variable, *i.e.*, the (weighted) mean of the thermal core and shell of the body, by measuring temperatures throughout the body and transmitting their data to a **controller**, located in the brain, where they are integrated and transduced into signals to **thermoeffectors** that are activated such that the temperature deviations caused by internal or external disturbances are suitably counteracted (see Chap. 6). A regulatory system that operates in this fashion is termed a **closed-loop negative feedback system** (see Chap. 3). The (weighted) mean body temperature, a deviation from which, up or down, activates these regulatory processes, is the so-called **set-point** of body temperature. The cenothermic T_c of *ca.* 37 °C is commonly used as the value of this set-point. But, as we have seen, it is represented in reality not by a single value, but a range of mean body temperatures.

B-8

6. Summary

The data presented indicate that the T_c of homeothermic animals is maintained within a narrow range which corresponds to the optimal temperature for the efficient operation of their bodily functions. However, the notion that 37 °C (98.6 °F) represents the "normal" T_c of adult humans should be relinquished since, in point of fact, this value does not correspond to the overall mean, the mean over a 24-h period, the median, or the mode for a large sample population; indeed, it lies outside the 99.9% confidence limits of that sample mean [2]. Furthermore, T_c is continuously influenced by many exogenous and endogenous variables (Table 2 and subsequent chapters).

The temperature of a tissue is a function of its rate of heat production and of the rate at which the heat generated flows away down a thermal gradient. Both these rates vary in different tissues, depending on numerous factors discussed in subsequent chapters. Consequently, local temperatures vary also, *i.e.*, the body has many different temperatures rather than one. Generally, T_c's are higher, more uniform, and stabler than T_{sk}'s. They are the ones that the body defends, *i.e.*, they are regulated.

7. References

[1] Aschoff J, Wever R (1958). Kern und Schale im Wärmehaushalt des Menschen. *Naturwiss.* 45: 477-485.

[2] Mackowiak PA, Wasserman SS, Levine MM (1992). A critical appraisal of 37 °C (98.6 °F), the upper limit of the normal body temperature, and other legacies of Carl Reinhold August Wunderlich. *JAMA* 268: 1578-1580.

[3] Rabinowitz RP, Cookson SL, Wasserman SS, Mackowiak PA (1996). Study of the effects of anatomic site, oral stimulation and body position on estimates of body temperature. *Arch. Intern. Med.* 156: 777-780.

[4] Hardy JD, DuBois EF (1938). Basal metabolism, radiation, convection and vaporization at temperatures of 22 to 35 °C. *J. Nutrition* 15: 477-497.

Addiditonal suggested reading:

Barnes RB (1963). Thermography of the human body. *Science* 140: 870-872.

Cooper KE, Veale WL, Malkinson TJ (1977). Measurement of body temperature. In: Myers RE (ed), *Methods in Psychobiology, vol. III*. Academic, New York, pp. 149-187.

DuBois EF (1951). The many different temperatures of the human body and its parts. *West. J. Surg.* 59: 476-479.

Edwards, RJ, Belyavin AJ, Harrison MN (1978). Core temperature measurement in man. *Aviat. Space Environ. Med.* 49: 1289-1294.

Fregly MJ, Blatteis CM (eds) (1996). *Handbook of Physiology, Sec. 4: Environmental Physiology, vol. I*. Oxford Univ. Press, New York.

Glossary of terms for thermal physiology, 2nd ed. (1987) *Pflügers Arch.* 410: 567-587.

Horvath SM, Menduke H, Piersol GM (1950). Oral and rectal temperatures of man. *JAMA* 144: 1562-1565.

Mackowiak PA (ed) (1997). *Fever: Basic mechanisms and management.* 2nd ed. Lippincott-Raven, Philadelphia.

Mead J, Bonarito CL (1949). Reliability of rectal tempertures as an index of internal body temperature. *J. Appl. Physiol.* 2: 97-109.

Silverman RW, Lomax P (1990). The measurement of temperature for thermoregulatory studies. In: Schönbaum E, Lomax P (eds), *Thermoregulation: Physiology and Biochemistry.* Pergamon, New York, pp. 51-60.

Self-study questions
1. What is body heat? How is it derived? Why is it important?
2. How can body heat be measured?
3. Is there really "one" body temperature?
4. How is "body" temperature measured?
5. What factors influence the thermal state of the body?
6. How is the core temperature maintained within relatively narrow ranges?
7. How is the thermoregulatory system organized?

Chapter 3

Learning objectives
1. Basic understanding of the physical processes of heat exchange: conduction, convection, radiation, evaporation.
2. Estimation and/or calculation of the amounts of the heat exchange components and of their relative contributions to total heat exchange.
3. Definition of parameters and variables used in heat transfer and temperature regulation.
4. Introduction to the formulation and use of the heat-balance equations of the body.
5. Basic properties of the thermal control loop.
6. Limitations of heat exchange and thermoregulatory control.

Bullets
- B-1 Double challenge of heat load.
- B-2 Heat loss is vital.
- B-3 Heat transfer within the body by conduction through the tissue and by convection via the blood.
- B-4 Bypass mechanisms.
- B-5 Countercurrent heat exchange.
- B-6 Heat flow due to perfusion.
- B-7 Simplified calculation of overall heat transfer.
- B-8 The still air layer is variable.
- B-9 Temperature difference and contact are essential for conductive heat transfer.
- B-10 The secret of clothing insulation consists of inclusion of air traps.
- B-11 An additional factor in convective heat transfer is wind speed.
- B-12 Radiant temperature is often different from air temperature.
- B-13 Easy estimation of "dry heat exchange".
- B-14 For evaporative heat loss differences of vapour pressures, not of temperatures, are essential.
- B-15 Respiratory evaporative heat loss is the main heat defence mechanism in many animals.
- B-16 In the cold, convection is the main component of heat loss; in the heat, it is the evaporation of sweat.
- B-17 Temperature is regulated within a closed control-loop with negative feedback.
- B-18 A control loop may be opened; it becomes a "passive system".

B-19 Heat balance is a necessary, but not a sufficient, condition for homeothermy.
B-20 The heat balance equations sum up heat gains and heat losses.
B-21 A block diagram elucidates the functional relationship of inputs and outputs of a system.
B-22 The uncontrolled thermal system follows external and internal disturbances.
B-23 The actuating signal of the controller is composed of weighted temperature signals core and mean skin temperatures may be considered as essential inputs.
B-24 Proportional control implies a load error.
B-25 Temperature differences (gradients) are essential for heat transfer.
B-26 Evaporated sweat generally is less than sweat secreted.
B-27 Evaporation is possible even into 100% water saturated air.
B-28 Net heat loss is possible even if skin temperature is lower than air temperature.

Chapter 3

BIOPHYSICS OF HEAT EXCHANGE BETWEEN BODY AND ENVIRONMENT

JÜRGEN WERNER
Ruhr-University, Medical Faculty, Center for Biomedical Methods
MA 4/59, D-44780 Bochum, Germany

3.1. Introduction

Figure 1 schematically shows a person producing heat within the exercising skeletal muscles. Heat is transported from the body core to the skin, and is subsequently exchanged with the environment via different physical processes explained in detail in section 3.2: conduction, convection, radiation, and evaporation. The exchange of body heat with the environment is modulated by air temperature and air humidity, wind speed, solar, sky and ground radiation, air pressure, posture and clothing. At the start of exercise, heat production exceeds heat dissipation, hence body heat storage occurs, causing an elevation of body temperature that evokes, via physiological control, heat loss mechanisms such as increased skin blood flow and volume as well as an increased sweating rate. In order to maintain thermal balance, an exercising person often has to cope with the double challenge of endogenous (metabolic) and exogenous (environmental) heat loads.

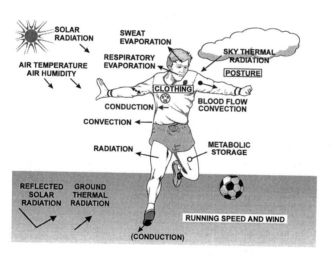

Fig. 1: Exercise and temperature regulation:
Heat production, heat transfer and environmental factors

The significance of heat exchange between body and environment becomes particularly evident in situations in which the body has to be thermally insulated from the environment for protection, e.g., from fire, ionizing radiation, or dangerous chemicals. In many of these cases, the body temperature has to be technically controlled by means of cooled or heated underwear, as is done, e.g., in astronauts' suits. Assuming perfect insulation, i.e., no heat exchange with the environment, body temperature would continuously increase because of continuous metabolic heat production. Without any heat loss, heat storage rate, S [in W], is calculated according to the general physical formula:

$$S = mc \frac{\Delta T_b}{\Delta t} \quad (1)$$

where mass, m, of the body may be 70 kg, and specific heat, $c = 3.47$ kJ kg^{-1} °C^{-1}. Assuming a minimal metabolic rate at rest of 80 W, we can calculate, using this formula, a critical increase of body temperature, ΔT_b, of 5°C, within a time period, Δt, of about 4 hours. Serious heat injuries may occur with a lesser increase of body temperature! Imagine that a jogger exercises at a speed of 3.3 m/s (about 12 km/h) at 70 % of his maximal oxygen uptake, which results in 2.8 L/min oxygen uptake, producing a 980 W metabolic rate. Without any heat loss, according to Equation (1) body temperature would rise by 5°C within only about 20 minutes !

Under comfortable conditions (meaning: about 30°C air and radiant temperature, unclothed body at rest in the early morning, low air humidity and velocity), heat loss to the environment is equivalent to internal heat production of the body, yielding a core temperature in a "standard" human of about 36.4°C and a mean skin temperature of about 33.7°C. If core temperature is weighted at 80 % and skin temperature at 20 % of the overall mean body temperature, T_b, this yields $T_b \approx 35.9$°C. In order to maintain body temperature near this value, heat exchange has to be actively controlled by the thermoregulatory centers. Thereby, in living organisms, heat exchange and its performance are determined by
1) the physical properties of the environment and of the body ("passive system")
2) the properties of the controlling systems
3) the properties of the closed control-loop.

Aspect 1) will be discussed in section 3.2, aspects 2) and 3) in section 3.3, and the limits of heat exchange in section 3.4.

3.2 Components of Heat Exchange
Metabolism produces heat which must be transported from the body core to the skin surface, where it is exchanged with the environment. Heat transfer within the body is achieved by conduction through the tissues and convection via the blood. Heat

exchange with the environment is accomplished by conduction, radiation, and evaporation. These physical processes and their underlying laws will be explained in the following sections.

3.2.1 Heat Exchange within the Body
3.2.1.1 Conduction through the Tissues

Physically, heat transfer through a non-moving material is called conduction. It follows a complex physical law, as in fact it is a continuous process with respect both to time and to local coordinates. However, for practical purposes, it is especially useful to know the net conductive heat flow, H_k [W], through an area, A, in a tissue layer of thickness, Δx (Fig. 2). This may be calculated easily, as it is inversely proportional to Δx and directly proportional to A, to the temperature difference at the layer boundaries, $(T_1 - T_2)$, and to a physical constant, the thermal conductivity, λ [W m^{-1} °C^{-1}], of the particular tissue:

$$H_k = \frac{\lambda A}{\Delta x}(T_1 - T_2) \qquad (2)$$

For values of λ see section 3.7, Table A.

Conduction may also occur as a transfer process from the body surface through still air, a water layer, clothing, or some other material or medium (see section 3.2.2.1).

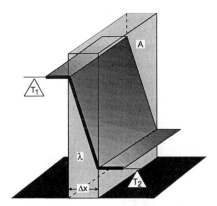

Fig. 2: Heat conduction through a solid wall of area, A, thickness, Δx, and thermal conductivity, λ. Surface temperatures, T_1 and T_2.

3.2.1.2 Convection via the Blood

Heat transfer via a moving gas or liquid is called convection, which is a very efficient component of heat transfer within the body, because it determines temperature differences within the body and controls the effective insulation in the skin region. Tissues having high metabolic rates are highly perfused by blood. By this mechanism, in spite of internal and/or external heat loads, temperature differences within the body are generally kept relatively small. Thus, cooler tissues are warmed by blood coming from heat-producing, active organs. Warm blood flow to the surface is increased by vasodilatation when the body needs to get rid of heat. Alternatively, blood can pass directly from arteries to veins via deeper channels through vasoconstriction of more superficial vessels, and particularly via arterio-venous anastomoses (shunts) when conservation of body heat is vital. These mechanisms raise or lower, respectively, the temperature gradient for heat transfer by conduction. Furthermore, arterial blood flowing along the body's extremities is precooled by loss of heat to adjacent venous streams (countercurrent heat exchange). This reduces the temperature of the limbs and lowers heat loss. Since most arteries lie deep, while veins occur both in superficial and deep regions, the extent of arterio-venous heat exchange depends on the route taken back to the body trunk by the venous blood. In the cold, large temperature differences exist between central and peripheral areas. See chap. 5 for further details.

B-4

B-5

The temperature of the arterial blood that enters the cutaneous plexi does not change much during its passage from the central arteries. However, on entering smaller cutaneous vertical vessels, the blood equilibrates very quickly with the local tissue temperature. Therefore, heat exchange via perfusion in the skin, the exact calculation of which is not possible on account of the complex circulatory architecture, may be described approximately by the so-called Pennes' bio-heat approach, considering capillary blood perfusion as spatially distributed heat sources or sinks. According to this approach volumetric heat flow, H' [W/m^3], is proportional to the product of volumetric blood flow, BF' ("perfusion"), and the difference between the arterial blood and tissue temperatures, (T_a- T):

$$H' = \rho_{bl} \cdot c_{bl} \cdot BF' \cdot (T_a - T) \qquad (3)$$

where ρ_{bl} is the density of blood and c_{bl} its specific heat.

This formula very often is used for calculation of the overall convective heat transfer in other types of tissue, but it has turned out to be rather inadequate, because, e.g., in muscles, heat exchange takes place predominantly in countercurrent arterio-venous trees with vessels of 0.2 - 0.5 mm diameter. Hence, in such cases, reliable calculations of heat transfer should be based on a vessel-by-vessel basis, which is a very intricate mathematical and computational task.

B-6

3.2.1.3 Overall Heat Transfer in the Body

For the purpose of assessing the overall heat transfer (conductive and convective), H, from core to skin and from skin to environment, it is convenient to use a tissue parameter which is not amenable to direct measurement, the so-called conductance, k [W m^{-2} °C^{-1}]. The inner (i) heat transfer from core (co) to skin (sk) is thereby calculated by:

$$H_i = k_i A_i (T_{co} - T_{sk}) \qquad (4)$$

and the outer (o) heat transfer from skin (sk) to the ambient surroundings (a) by:

$$H_o = k_o A_0 (T_{sk} - T_a) \qquad (5)$$

Also, using total heat conductance, k_{tot}, heat transfer through the whole body can be determined:

$$H_{tot} = k_{tot} A (T_{co} - T_a) \qquad (6)$$

If the body is in thermal steady state, i.e., no temperature changes occur, there is no difference between inner, outer and total heat transfer.

The inverse values of heat conductances are named thermal tissue resistances,

$$R_i = 1/k_i \quad \text{and} \quad R_o = 1/k_o, \qquad (7)$$

respectively. For the clothing layer, the thermal resistance is termed thermal insulation, I.

For the purpose of calculating the total heat balance for body compartments (e.g., core, skin) or the whole body, it is convenient to define absolute conductances, k*[W/°C], or resistances, R* [°C/W]:

$$k^* = k \cdot A \qquad (8)$$

$$R^* = R/A = 1/kA = 1/k^* \qquad (9)$$

Heat transfer, e.g., between the body core (co) and the skin (sk), may thus be summarized as follows:

$$H = k^*(T_{co} - T_{sk}) \qquad (10)$$

3.2.2 "Dry" Heat Exchange with the Environment

"Dry" heat exchange with the environment is based on conduction, convection, and radiation. Each of these components may contribute to either heat loss or heat gain (see 3.3.1 and 3.4.).

3.2.2.1 Conductive Heat Exchange (K)

As conduction is defined as heat transfer through a non-moving material (remember section, 3.2.1.1), purely conductive heat exchange with the environment occurs when the body or parts of it are in close contact with solid materials. If not in contact, even with the body and environmental air immobile, so-called free convection develops (see below). Even at low wind speeds, the human body is surrounded by a boundary layer of virtually still air which may have a thickness between 4 and 8 mm. A wind velocity of 2 m/s reduces this layer thickness to about 1 mm. Conductive heat exchange also occurs between body and clothing.   **B-8**

Conductive heat transfer, K, is proportional to the difference of skin and ambient temperatures, and to the contact area, A_k, using the conductive heat transfer coefficient, h_k [Wm^{-2}°C^{-1}]:   **B-9**

$$K = h_k A_k (T_{sk} - T_a) \tag{11}$$

The amount of purely conductive heat loss to objects may be, depending on their heat capacity, of minor importance because, if the object heats up, it does not thereafter provide much of a heat sink. In fact, the object may then act more as insulation as it prevents heat loss by other modes. However, in the case of high and low temperatures, direct contact of skin areas to objects may be very harmful, e.g., by causing burns or frostbites, respectively.

The heat transfer coefficient of clothing depends primarily on the ability of the material to include air traps (e.g., wool). In water, conductive heat loss is very much higher than in air as the specific heat of water is 4000 times that of air, and its thermal conductivity around 25 times greater (cf. Table A in section 3.7).   **B-10**

3.2.2.2 Convective Heat Exchange (C)

If heat transfer occurs via moving gas or liquid, it is called convection (remember section 3.2.1.2). However, so-called free convection develops even in still air, because the temperature differences between body and environment induce air movement. Forced convective heat exchange develops in the presence of ambient air or water flowing outside the conductive air layer mentioned in the preceding paragraph. For practical purposes, the combined components including the conductive process in the small air layer are summarized in an equation analogous to Equation (11):

$$C = h_c A_c (T_{sk} - T_a) \tag{12}$$

The combined convective heat transfer coefficient, h_c, being dependent primarily on the shape of the body surface (radius, position, such as standing or sitting), on air or water velocity, v, and on the direction of flow (also on air pressure, if it changes substantially), may be written as:

B-11

$$h_c = c_1 v^{c_2} \tag{13}$$

In air, c_1 may vary from about 2 to 10 Wsm^{-3}°C^{-1}, and c_2 from about 0.5 to 0.7.

3.2.2.3 Radiative Heat Exchange (R)

Even if there is vacuum between two bodies which have different temperatures, we observe heat transfer, called radiation. It is an ubiquitous process in addition to conduction and convection. Radiative heat transfer may be present in the form of solar radiation, sky thermal radiation, reflected solar radiation, ground thermal radiation, and of radiation to or from any objects. Radiation from surrounding walls is often underestimated. According to the Stefan-Boltzmann-law, the radiative heat transfer component, R, is:

$$R = \sigma \cdot \varepsilon \cdot A_r (T_{sk}^4 - T_r^4) \tag{14}$$

where σ is the Stefan-Boltzmann radiation constant, ε the emissivity of the skin surface, and T skin (sk) and radiant (r) temperatures measured in Kelvin [K]. For infrared radiation, human skin has practically the emissivity of an ideal radiator, the so-called "black body", $\varepsilon \approx 1$, independent of skin colour. (However, sun radiation is, in contrast to infrared radiation, absorbed according to the colour, the absorption coefficient lying between 0.55 and 0.85).

B-12

For small temperature differences, it is not necessary to calculate the 4th power of the temperatures. A mathematical linearization delivers a combined proportionality constant, the so-called radiative heat transfer coeficient, h_r, with values around 4 Wm^{-2}°C^{-1}, depending on posture. Thus, instead of Equation (14), a simple linear relation, analogous to Equations (11) and (12) is used:

$$R = h_r A_r (T_{sk} - T_r) \tag{15}$$

For purposes of simple estimations, often the convective and the radiative coefficients are summed up in the "dry" transfer coefficient, h. However, if $T_a \neq T_r$, a weighted average of air and radiative temperatures is used, the so-called operative temperature, T_o, in order to be able to combine Equations (12) and (15).

$$C + R = hA (T_{sk} - T_o) \tag{16}$$

B-13

3.2.3 Heat Loss by Evaporation of Sweat (E)

Any transition from the liquid to the gaseous state needs energy. This is taken from the surface where this process takes place. If it occurs below the boiling point, it is called evaporation. Evaporation of sweat induces heat loss from the body. So it should be kept in mind that only that portion of sweat which is evaporated results in heat loss from the body. It has to be taken into account that part of the sweat will be lost as droplets dripping from the body before evaporation. This is particularly true in exercising or working individuals. In this paragraph, we assume that enough sweat is available for evaporation. If the total sweat rate, SWR, is evaporated, the evaporative heat loss may be calculated, using the latent heat of vaporization, LHV:

$$E = LHV \cdot SWR \tag{17}$$

However, the maximal evaporation cannot be greater than the amount determined by the difference of the vapour pressures on the skin surface and in the environment and the wetted area, A_{wet}:

$$E = h_e A_{Wet} \left[p(T_{sk}) - p(T_a) \right] \tag{18}$$ **B-14**

with the evaporation transfer coefficient, h_e, being dependent primarily on the form of the body surface and on air velocity (and on air pressure):

$$h_e = e_1 v^{e_2} \tag{19}$$

with $e_1 = 0.4 - 20$ Wsm^{-3}mmHg^{-1} and $e_2 = 0.4 - 0.7$. Remember that heat loss due to sweat secretion means both water loss and loss of electrolytes from the body (see chapter 5)!

3.2.2.4 Respiratory Evaporative Heat Loss (E_R)

Evaporation also takes place in the respiratory tract. In humans, this route of thermoregulatory heat loss is of minor relevance, whereas most mammals and birds effectively amplify, via high rates of ventilation ("panting"), respiratory evaporative heat loss. For a quick assessment of the amount of E_R in exercising humans, the following formula may be used:

$$E_R = 0.0023M \left[(44 - p_a) + 0.61(35 - T_a) \right] \tag{20}$$

This equation assumes that the expired air temperature of exercising subjects remains essentially constant at 35°C over a wide range of ambient conditions and that the water vapour pressure of expired air stays at 44 mmHg. p_a is the water vapour pressure of the environment in mmHg, T_a is air temperature in °C, and M is metabolic rate (= rate of metabolic energy transformation). In resting subjects, amounts up to 12 W may be lost. **B-15** According to Equation (20), in exercising subjects, E_R may be increased tenfold or more over resting values. However, the assumptions made in Equation (20) are not valid in a very cold environment.

3.2.5 Contribution of the Components of Heat Loss

Summarizing all modes of heat exchange discussed, we get the classical heat balance equation for the whole body, stating that heat storage, S, occurs, if there is an imbalance of metabolic rate, M, and the heat exchange components outlined above:

$$S = M \pm K \pm C \pm R - E - E_R \tag{21}$$

(Problems of heat balance will be further discussed in sections 3.3, 3.3.1, and 3.4).

The quantitative contribution of the components of heat loss in steady state is demonstrated by the following, simple example. If a jogger exercises at 70 % $V_{O_2 max}$ (2.8 L/min oxygen uptake assumed) while wearing shorts and running at a speed of 3.3 m/s (about 12 km/h) in the sun, he may have a 980 W metabolic rate and a 140 W heat gain by solar radiation; thus the body has to accommodate a total heat gain of 1120 W. Mechanical power is considered to be negligible in this example. Figure 3 depicts the four components of heat loss, C, R, E, and E_R, in an environment with a 60 % relative humidity, RH, and ambient temperatures from 10°C to 30°C. At low air temperatures, convective heat loss is the dominant component, and evaporative heat loss is relatively low. In a warm environment, evaporation becomes the essential mode of heat loss.

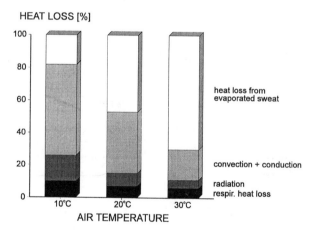

Fig. 3: Components of heat loss during exercise (jogging 2.8 L/min oxygen uptake, 140 W solar radiation, 60 % relative humidity) at different ambient temperatures. Data from Mitchell

Radiative and respiratory heat losses remain relatively low, with a tendency to fall when ambient temperature is higher. Humidity effects on heat loss are apparent in

Equation (18) via the effects of vapour pressure. In the example of a jogger at $T_a = 30°C$ and RH = 60%, the wetted skin area, A_w, is only 0.4; at RH = 80%, A_{wet} would rise to 0.55, whereas at RH = 20% A_{wet} would drop to 0.25. Even at 100 % RH, the skin wetness would be only 75%; thus, in even such a humid environment, the runner could still dissipate heat during exercise.

3.3. Basic Properties of the Thermal Control Loop and its Subsystems

Heat exchange in the human body takes place within a closed control loop. This means that body temperatures are continuously measured by thermoreceptors and fed back via neuronal signals to the central nervous system, activating effector mechanisms with the aim of counteracting internal or external disturbances, e.g., due to influences of muscular exercise and of the external climate, respectively. For an understanding of the function of such closed-loop systems, it is convenient to open the loop fictitiously and to analyze the open-loop characteristics of the subsystems, i.e., the controlling system and the system which is controlled. The latter, considered as an open-loop system, is the so-called "passive system" which, without control, would passively follow the external and internal disturbances. It is characterized by the physical laws of heat transfer and exchange.

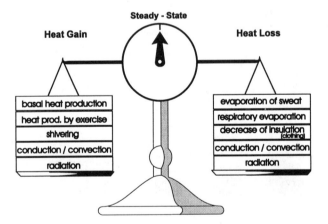

Fig. 4: Balance of heat gain and heat loss guarantees steady state (net heat flow = zero, i.e., constant body temperature). However, such a steady state does not imply that body temperature is equal to that value which prevails in a resting individual under comfortable environmental conditions. In addition, if there is any imbalance of heat gain and heat loss, body temperature will start to increase or decrease, respectively.

Each internal and external disturbance will evoke heat transfer processes in the passive system, whether controlled or not. As long as the temperatures are compatible with human life and as long as the degree of disturbance does not change, even the passive system will achieve a steady state determined by the heat balance of the body (Fig. 4), whether controlled or not. Hence, it is not only the state of heat balance per se which is achieved by feedback control but that level of heat balance which, in spite of internal or external heat loads, enables body temperature to approximate "normal" body temperatures, i.e., those values which would prevail also without the activation of heat or cold defence. Heat balance is a necessary, but not a sufficient condition for homeothermy!

3.3.1 The Heat Balance Equations for the Passive System

Metabolism produces heat which has to be transported from the body core to the skin surface where it is exchanged with the environment. Therefore, it is reasonable and, in most cases, sufficient to formulate the heat balance equation for two compartments, core (co = bones, muscles, viscera) and skin (sk; includes fat layer). Heat flow or energy storage, S, is the product of mass, m, specific heat, c, and change of temperature with respect to time, dT/dt. The heat balance equations state that overall heat flow equals the difference between heat gain and heat loss. For the core, heat gain is identical to metabolic rate, M, minus or plus (if exercise with eccentric muscle actions is performed) mechanical power, W, whereas heat loss is generally composed of conductive/convective heat transfer, H, to the skin, and heat loss in the respiratory tract, E_R. Note that several terms introduced as heat loss components may, under the conditions defined in 3.4, change their sign, i.e., turn into heat gain:

$$m_{co} c_{co} \frac{dT_{co}}{dt} = M - W - H - E_R \qquad (22)$$

Normally the skin compartment receives the conductive/convective heat transfer from the core, H, as heat gained, whereas evaporative heat loss, E, and heat transferred to the environment by conduction and convection, C, and by radiation, R, constitute the heat loss components:

$$m_{sk} c_{sk} \frac{dT_{sk}}{dt} = H - E - C - R \qquad (23)$$

H is defined by Equation (10) and (C+R) by Equation (16). They are set into Equation (22) and (23) assuming $T_o = T_a$:

$$m_{co} c_{co} \frac{dT_{co}}{dt} = M - W - k^*(T_{co} - T_{sk}) - E_R \qquad (24)$$

$$m_{sk} c_{sk} \frac{dT_{sk}}{dt} = K^*(T_{co} - T_{sk}) - E - hA(T_{sk} - T_a) \qquad (25)$$

M is composed of basal heat production, M_0, heat production due to shivering, ΔMSH, and due to exercise, ΔMEX. Using these definitions, in Fig. 5, Equations (24) and (25) are transformed directly into a block diagram.

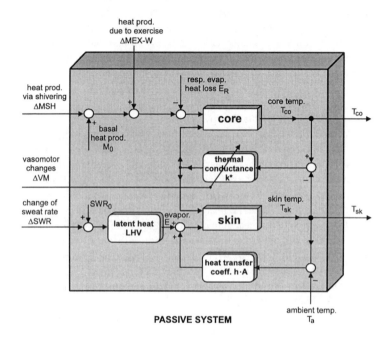

Fig. 5: Functional relationship of the passive thermal system (body as an uncontrolled open-loop heat transfer system). Output of the system: core temperature, T_{co}, and skin temperature, T_{sk}. Disturbance to the system by an ambient temperature, T_a, and exercise. Possible control input by heat production via shivering, ΔMSH, change of sweat rate, ΔSWR, and vasomotor changes, ΔVM.

It shows
1) T_{co} and T_{sk} as output information from the passive system.
2) Heat production due to exercise reduced by mechanical power, ΔMEX-W, and ambient temperature, T_a, as the main disturbing factors.
3) Shivering, ΔMSH, change of sweat rate, ΔSWR, and vasomotor changes, ΔVM (evident by alterations in tissue conductance, k^*), as possible sites of impact of controlling mechanisms.

Total evaporative heat loss is determined by multiplying evaporated sweat with latent heat of vaporization, LHV. Fig. 5 further shows that, via the tissue conductance, k^*, and via the heat transfer skin/environment transfer coefficient, h, the passive, unregulated system already possesses two inherent feedback loops which, however, do not constitute a temperature control loop.

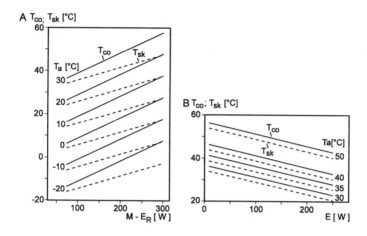

Fig. 6: Characteristics of the passive thermal system for core temperature, T_{co}, and skin temperature, T_{sk}. Vasomotor change is not taken into account. (A) T_{co} and T_{sk} as a function of metabolic heat production, M, minus respiratory heat loss, E_R; (B) T_{co} and T_{sk} as a function of evaporative heat loss, E, via the skin. T_a = ambient temperature.

If the thermal system is in steady state, there are no further temperature changes, dT/dt (i.e., no further heat storage), and Equations (24 and 25) are equal to zero. By solving these equations for T_{co} and T_{sk}, we obtain the following steady state characteristics of the passive system:

$$T_{co} = \frac{k^* + hA}{k \cdot hA}(M - W - E_R) - \frac{1}{hA}E + T_a \tag{26}$$

$$T_{sk} = \frac{1}{hA}(M - W - E - E_R) + T_a \tag{27}$$

T_{co} and T_{sk} are shown in Fig. 6A as a function of M - E_R, and in Fig. 6B as a function of E. The lines represent the open-loop characteristics of the unregulated body core and shell, provided that heat resistances do not change, i.e., that there are no vasomotor changes. They show, assuming a constant metabolic rate, that the body temperature decreases with decreasing ambient temperature and, assuming a constant evaporative rate, increases with increasing ambient temperature. Changing vasomotor activity changes the total heat resistance, i.e., the slope of these characteristics.

3.3.2 Principles of the Controlling System and of Negative Feedback Control

In order to maintain normal body temperature, T_{co} and T_{sk} are sensed by thermoreceptors and fed back as afferent action potential trains to the central nervous system (Fig. 7; see also chap. 6). The simplest algorithm, for which much experimental evidence has been gathered, assumes that the essential integrated signal is formed by a weighting of core and skin temperatures with a weighting factor, g:

$$a = g\, T_{co} + (1 - g)\, T_{sk} \tag{28}$$

which may be interpreted as the weighted central information on body temperature, T_b. If this signal is above or below a threshold, a_0, the effector mechanism, ΔEFF, will be activated, the value being determined by the gain, G, of the controlling system, the value of which is, of course, different for each of the effector systems:

$$\Delta EFF = G \cdot (a - a_0) \tag{29}$$

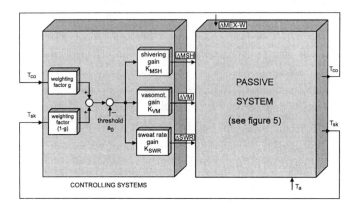

Fig. 7: Combining the passive system from Fig. 5 via negative feedback with controlling systems guarantees a body temperature near to the value in the indifferent zone (without disturbances). ΔMSH = heat production via shivering, ΔVM = vasomotor changes, ΔSWR = change of sweat rate, ΔMEX-W = heat production due to exercise minus mechanical power, T_a = ambient temperature, T_{co} = core temperature, T_{sk} = skin temperature

The difference $(a - a_0)$ may be interpreted as the "load error", ΔT_b. Fig. 7 outlines this basic concept for the three effector mechanisms, viz., shivering, ΔMSH, vasomotor changes, ΔVM, and sweat rate, ΔSWR. (In more elaborate concepts, different thresholds for different effector mechanisms may be taken into account.) By this process, negative feedback is achieved which, to a large extent, compensates for the influences of the internal disturbance (heat production due to exercise) and the external climatic factors. However, the outlined control algorithm (Eq. (29), so-called proportional control) implies, when disturbances are present, a permanent load error, meaning that, e.g., the influences of ambient temperature are drastically reduced, but that nevertheless body temperature in the heat will be above, and in the cold will be below the value observed in thermoneutral conditions. This fact, often not understood even by experts in thermal physiology, can be explained as follows: let us, e.g., consider the case of heat defence by sweat production (for the sake of simplicity, we will not consider associated vasomotor changes):

Changes of ambient temperature, ΔT_a, and changes of evaporative rate, ΔE, will evoke body temperature changes, ΔT_{co} and ΔT_{sk}. Formulating Equations (26) and (27) in terms of such deviations, Δ, all components which do not change will vanish (M, W, E_R). Hence, we get an extremely simple equation valid for ΔT_{co}, and ΔT_{sk}, but also for change of body temperature, ΔT_b:

$$\Delta T_b = -(1/hA) \cdot \Delta E + \Delta T_a \tag{30}$$

The general effector equation (29) is formulated for proportional control of sweat rate, ΔSWR, hence of evaporative rate, ΔE:

$$\Delta E = LHV \cdot \Delta SWR \tag{31a}$$

$$\Delta E = LHV \cdot G \cdot \Delta T_b \tag{31b}$$

where LHV, again, is the latent heat of vaporization. The control loop represented by these two simple equations is drawn in Fig. 8.

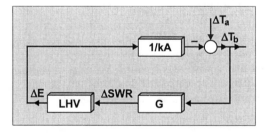

Fig. 8: Proportional control in heat defence. ΔT_a = change of ambient temperature, ΔT_b = change of body temperature, ΔSWR = change of sweat rate, ΔE = change of evaporative heat loss, LHV = latent heat of vaporization, G = gain factor of sweat controller, 1/hA = thermal resistance of the skin. Note the minus sign in the control loop which guarantees feedback control. However, proportional control implies a load error, i.e., deviations of body temperature are minimized, but not zero.

Setting Eq. (31), which describes the passive system, into Eq. (32), which describes the control system, we get one equation for the closed control loop, relating the output, deviation of body temperature ("load error"), ΔT_b, to the input, change of ambient temperature, ΔT_a. The parameters h, A, LHV and G may be combined in the so-called control-factor α:

$$\Delta T_b = \frac{1}{1 + (1/hA) \cdot LHV \cdot G} \cdot \Delta T_a \qquad (32a)$$

$$\Delta T_b = \alpha \cdot \Delta T_a \qquad (32b)$$

Thus, a disturbance in ambient temperature (T_a), i.e., a deviation of T_a, ΔT_a, from the indifferent state or "comfort zone" would be expected to cause a deviation of body temperature or load error, ΔT_b (Eq. 32a). The occurrence of this load error is inevitable since it is inherent in the concept of "proportional control" [see Eq.(29)]. The actual magnitude of ΔT_b, however, would depend [see Eq. (32b)] on the control factor, α, which [see Eq. (32a)] would be low if the controller gain, G, were high. Hence, a large controller gain infers a small ΔT_b; or, put another way, a high degree of effector responsiveness serves to reduce the effect of the original disturbance in T_a on T_b. Only in the theoretical case that G were raised to infinity would α and , hence, the load error be 0.

Despite of the permanent load error, the deviation of body temperature in proportional control is very much lower than in a system without control ("passive" system). From Eq. (30) it is evident that, without control, ($\Delta E = 0$), deviation of body temperature, ΔT_b, would be identical to ΔT_a. This would be much greater than in the case of closed-loop control, where ΔT_a multiplied by a control factor $\alpha \ll 1$, yields ΔT_b.

3.4. Limitations of Heat Exchange

If the level of metabolic heat production or the environmental load changes, all heat loss components as shown in Eq. (22) - (25) will also change because they depend on the accompanying changes in core and/or skin temperatures. Heat transfer between core and skin is proportional to the temperature difference between core and skin: see Eq. (10).

As defined in the heat balance equations, (22) - (23), a positive value of H constitutes heat loss from the core. This is obtained only if Eq. (10) has a positive sign, and therefore if:

$$T_{co} > T_{sk} \qquad (33)$$

it is obvious that respiratory heat loss is limited, too.

Convective heat loss from the skin to the environment is proportional to the temperature difference between skin and air (Eq. 12). Hence, convective heat loss is limited to conditions in which the temperature of the skin is greater than that of the environment:

$$T_{sk} > T_a \tag{34}$$

It may be enhanced by increasing skin temperature (to increase the skin/environment gradient), the area of exposed skin, or the air velocity, which increases the heat transfer coefficient. Radiative heat loss may be written in an analogous way to convective heat loss (Eq. 15), with basically the same limitations and possibilities for enhancement.

Evaporative heat loss from the skin, Eq. (18), depends on the availability of sweat. However, the evaporated sweat, SWR_{ev}, is generally lower than the rate of sweat production because of droplet formation and because the upper limit of evaporation depends on the difference between the vapour pressures in the air, $p(T_a)$, and on the skin, $p(T_{sk})$. The non-linear physical relationship between water vapour pressures, relative humidity and air temperatures is shown in Fig. 9.

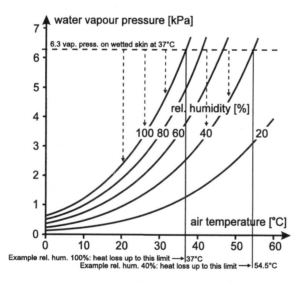

Fig. 9: Physical relationships between water vapour pressure, air temperature and relative humidity. For explanations of examples, see text.

Saturated vapour pressure is reached at 100 % relative humidity. It increases with increasing temperature. The vapour pressure on wetted skin at 37°C is 6.3 kPa (≙ 47 mmHg): see horizontal line in Fig. 9. Any state of the environment characterized by a combination of air temperature and relative humidity which, in the diagram, is located below this line allows for evaporation. Positive differences of the vapour pressures of skin and air are indicated by broken vertical lines. Thus, it is evident that evaporation is even possible into 100 % saturated air as long as air temperature is lower than the elevated skin temperature (37°C). If relative humidity, RH, is low, evaporation will be possible up to high air temperatures, T_a (see 2nd example in Fig. 9: RH = at 40 %, up to about 54°C). If T_a is held constant, maximal evaporation increases with decreasing RH. With constant RH, maximal evaporation would increase with decreasing T_a. With constant water vapour pressure of the air, various combinations of RH and T_a make possible identical maximal evaporative rates, higher T_a require lower RH, and vice versa.

If we substitute in the equations the vapour pressures, p, with relative humidities, RH, and saturated vapour pressures, p_{sat}, we get:

$$p(T_{sk}) - p(T_a) = RH_{sk}\, p_{sat}(T_{sk}) - RH_a\, p_{sat}(T_a) \tag{35}$$

With wetted skin and $RH_{sk} = 1$, this expression guarantees heat loss:

$$p_{sat}(T_{sk}) > RH_a\, p_{sat}(T_a) \tag{36}$$

which may be achieved by lowering RH_a or by increasing T_{sk}.

Assuming that the exposed skin areas for convective and radiative heat exchange are the same (=A) and that air and radiative temperatures are identical, and setting h = h_c + h_r and $h_e \approx 2.2\, h_c$, the total possible heat loss, HL, from the skin may be written simply as:

$$HL = hA(T_{sk} - T_a) + 2.2 h_c\, A_{wet}\, [p(T_{sk}) - p(T_a)] \tag{37}$$

showing very clearly the main limitations. These are:

a) environmental factors limiting either the transfer coefficients at low air velocity or the evaporative heat loss at high vapour pressure of the air (high relative humidity) and

b) regulatory factors such as limitations of blood flow distribution, resulting in a skin temperature lower than air temperature (or a lower vapour pressure at the skin), or limitations of sweat production, resulting in a wetted area less than the total body surface area.

Using Equations (17), (18), and (19), HL can be rewritten as:

$$HL = hA \left[T_{sk} - \left(T_a - \frac{SWR_{ev} \cdot LHV}{hA} \right) \right] \quad (38)$$

showing that heat loss is guaranteed as long as:

$$T_{sk} > T_a - \frac{SWR_{ev} \cdot LHV}{hA} = T_a^* \quad (39)$$

Thus, net heat loss is possible if skin temperature is lower than air temperature, but T_{sk} must be higher than the above defined temperature T_a^*.

3.5 Summary

Heat exchange and its performance are determined by the physical properties of the environment and of the body ("passive system"), by the properties of the controlling systems, and by the properties of the closed control loop.

Heat exchange within the body takes place by conduction and convection. Convection via a single vessel depends mainly on the vessel diameter and the blood flow velocity. The overall convective effect in microvessels may be considered to be proportional to perfusion rates and to the difference of arterial and tissue temperatures.

Heat exchange with the environment occurs by means of conduction, convection, radiation and evaporation. The overall "dry" heat loss is proportional to the body surface and the difference between skin and ambient temperatures, whereas heat loss by evaporation is proportional to the difference of water vapour pressures of skin and air. Heat transfer coefficients depend mainly on air/water velocity.

The body, if it were an uncontrolled heat transfer system, would passively follow the changes of environmental temperature. However, the controlling subsystems, thermoreceptors, integrating centers and effector mechanisms such as vasomotor reactions, shivering and sweat production, constitute, together with the passive system, a closed control loop with negative feedback. This prevents mean body temperature from deviating extensively from this value taken under thermoneutral conditions, in spite of external and/or internal thermal loads. This ensures optimal thermal conditions for all internal processes. With external or internal heat loads, mean body temperature will, in spite of the closed-loop control, deviate to an extent proportional to the thermal stress. This is an inherent property of every proportional controller, in which the amount of effector action is proportional to the deviation of the controlled variable. The limitations of thermoregulatory control are due to inherent effector capacities and to environmental factors limiting heat transfer and heat exchange.

3.6 Suggested Readings
[1] Gagge APH, Gonzales RR (1996). Mechanisms of heat exchange: biophysics and physiology. In: Fregly MJ, Blatteis CM (eds.), *Environmental Physiology*. Oxford University Press, New York - Oxford, pp. 45-84.
[2] Mitchell JW (1977). Energy exchanges during exercise. In: Nadel ER (ed), *Problems with Temperature Regulation during Exercise*. Academic Press Inc., New York, pp. 11-26.
[3] Werner J (1990). Functional mechanisms of temperature regulation, adaptation and fever. In: Schönbaum E and Lomax P (eds), *Thermoregulation, Physiology and Biochemistry*. Pergamon Press, New York, pp. 185-208.
[4] Werner J (1996). Modeling responses to heat and cold. In: Fregly MJ, Blatteis CM (eds.), *Environmental Physiology*. Oxford University Press, New York - Oxford, pp. 613-626.

3.7 Appendix
Table A: Thermal conductivity of normal unperfused tissue and some other materials

Tissue/ Material	Temperature [°C]	Thermal Conductivity [$W\ m^{-1}\ °C^{-1}$]
Blood	37°	~ 0.5
Muscle	37	~ 0.5
Fat	37	~ 0.2 ... 0.3
Skin	37	~ 0.3 ... 0.4
Air	20	~ 0.025
Water	20	~ 0.6
Platinum	20	~ 0.7
Copper	20	~ 0.4
Wood	20	~ 0.1 ... 0.2
Wool	20	~ 0.04

Self-evaluation questions for chapter 3

1. What would be the estimated increase of mean body temperature after about 20 minutes of jogging at 70 % of $V_{O_2 max}$, if heat loss to the environment were totally prevented?

2. What mechanisms account for the fact, that in neutral and warm conditions, temperature differences within the body remain relatively small?

3. How may heat flow due to microcirculatory perfusion be estimated?

4. What is the definition of total heat conductance?

5. How do you calculate conductive heat exchange?

6. How do you calculate convective heat transfer to or from the environment?

7. How do you calculate radiative heat exchange?

8. How do you calculate evaporative heat loss?

9. To what extent do the components of heat loss contribute to the overall heat loss, depending on environmental temperature?

10. What are the components of the heat balance equations of the body?

11. What is meant by the term "passive thermal system"?

12. How is negative feedback control achieved in thermoregulation?

13. What are the main physiological effector mechanisms?

14. Why does the thermoregulatory system exhibit a "load error"?

15. What main factors limit thermoregulatory control?

Chapter 4.1

Learning Objectives
1. Basic understanding of the thermoregulatory significance of shivering.
2. Definition of variables inducing shivering.
3. Definition of neural components regulating shivering.
4. Estimation of the metabolic efficiency of shivering.

Bullets

B-1 Shivering is a prerequisite of homeothermy.
B-2 Shivering occurs in almost all muscles of the body.
B-3 Shivering is thermogenically uneconomical.
B-4 Shivering is a nervious reflex.

Chapter 4.1

SHIVERING

LADISLAV JANSKÝ
Faculty of Science, Charles University
Prague, Czech Republic

1. **Introduction**

The „Glossary of Terms for Thermal Physiology" [1] describes shivering as „involuntary tremor of skeletal muscles as a thermoeffector activity for increasing metabolic heat production". Shivering can be also defined as „an increase in reflex, nonlocomotor muscular tone attributable to exposure to cold, with and without visible tremor" [2]. Jung [3] described shivering as „rhytmic involuntary movements of muscles, consisting of an oscillation about the midpoint of one or several limbs, the movements not causing any change in body position". Shivering appears as random, uncoordinated, simultaneous contractions of flexor and extensor muscles. This is different from the contractions occurring in parkinsonian, cerebellar, or clonic tremors. The subject of shivering has been rewieved recently by Kleinebeckel and Klussmann [2].

2. **Thermoregulatory significance of shivering**

Shivering and muscle tone are considered to be the most important sources of heat production in cold-exposed homeotherms. Shivering is a necessary prerequisite of homeothermy - species not exhibiting shivering can be considered as primitive from the point of evolution. In most species, shivering results in an increase in metabolism of two-three fold over the basal rate. In humans, an increase up to five times, or even higher has been reported. In rodents, this additional heat production enables the animals to withstand a temperature gradient between the body and the environment of 50°C or greater. In newborns and cold-adapted animals, shivering represents a second line of defence against cold. Mild cold stimuli evoke nonshivering thermogenesis, while stronger cold stimuli additionally induce shivering.

3. Localization of shivering

Muscular tremor in humans may appear as soon as after 2 min of cold exposure and becomes generalized in 24 min, depending on ambient and/or core temperature (see below). It involves almost all the muscles of the body, with the exception of the facial, perineal, extraocular, and middle ear muscles. In rabbits, the greatest intensity was measured in the neck and the lowest in the legs.

Shivering is a discontinuous process and fluctuates between different areas of the body. Thus, it appears first in the masseters and then extends to other muscle groups. There is controversy as to whether it occurs first at the periphery and then extends to the trunk, or whether it starts in the centre of the body and then extends to the legs.

4. Recordings of shivering

Electromyographic recordings are usually used to measure shivering. However, present techniques are far from adequate, since shivering is recorded only from a small area of the muscle, and little information concerning the interaction of individual muscles and muscle fibers in different parts of the body is obtained by this technique.

Shivering is normally recorded from an area of muscle, rather than fom a single motor unit, and summated electrical activity from several muscle units or even muscles is evaluated as peak-to-peak differences or is integrated. It has been shown that in rats the amplitude of shivering increases from 20 to 70 µV with cold stimulus. Frequency of shivering depends on the body size, being higher (50Hz) in smaller species than in bigger ones (15Hz) (Fig.1).

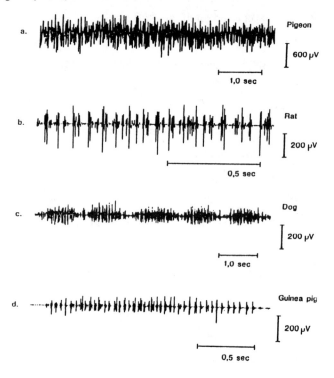

Fig.1 Examples of typical EMG recordings during shivering in the bird and mammal: /a/ continous activity from the pectoralis major muscle of the pigeon, /b/ grouped discharges from the tibialis anterior muscle of the rat, /c/ burst-like activity with several bursts from a thigh muscle of the dog, /d/ burst-like activity (single burst) from a thigh muscle of the guinea pig (from Kleinbeckel and Klussmann [2]).

Power spectral analysis indicate that in rabbits the spectrum of action potential frequencies is not changed by cold stress, the range of maximum frequencies being between 50 and 75 Hz.

5. Manifestation of shivering

It has been documented that slight cooling of the whole body first induces activity in resting muscles described as „thermal muscle tone". This tonic state of contraction appears earlier in extensor muscles than in flexors. It can be detected as fine electrical discharges with an amplitude of about 5 µV, the frequency ranging between 5 and 12 Hz. There is no synchronous firing of motor units within the same or antagonistic muscles. The units fire at different frequencies and out of phase with each other. The more intensive cooling evokes greater muscle oscillations. The amplitude of firing rate increases and the discharges tend to become synchronous, so that group discharges („bursts") corresponding to the contraction phase of each shivering cycle appear, followed by complete silence in the relaxed phase (Fig.2). The coordination of various motor units during this process is due to periodic inhibition of impulses in afferent muscle fibers, i. e., due to an oscillatory character of the motor nerve-afferent arc system, rather than by the central synchronization of these impulses. Muscle spindle proprioceptors are involved in this process [4]. Recurrent inhibition through Renshaw cells may be responsible not only for synchronization of motoneurones during shivering, but also for the frequency of shivering. More vigorous shivering results from addition of more muscle units to the synchronized population.

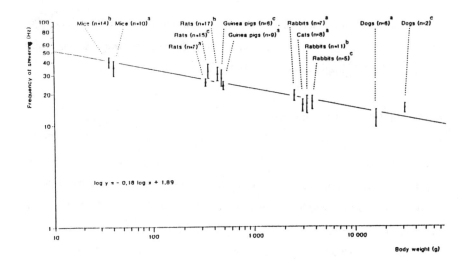

Fig. 2. Correlation between frequency of shivering and body weight in various adult mammals of different body size. Double logarithmic scale. (from Kleinebeckel and Klussmann [2]). The correlation coefficient /r/ of the regression line is 0.94.

6. Metabolic aspects of shivering

The energy for all muscle activity is generated from ATP. Contractile proteins in the muscle convert ATP to ADP, inorganic phosphate and energy. Since there is only a limited amount of ATP in muscle (3 mmol . kg $^{-1}$, which corresponds to the energy required for about eight contractions), ATP must be quickly regenerated from ADP. Thus, availability of ADP becomes a crucial factor responsible for determining the intensity of oxidative phosphorylation. The rate of respiration during shivering is regulated by a phosphorylation state ratio expressed as ATP/ADP : P.

Ion pumping also contributes to the increase in heat production during muscle activity, at least to a minor extent. The restoration of the normal polarized state of the sarcolemma involves pumping of ions by the Na+/K+ - activated ATPase. Relaxation requires the removal of Ca^{2+} from the cytosol by the Ca^{2+} - ATPase of the sarcoplasmic reticulum. Both of these processes accelerate metabolism via changes in the phosphorylation state ratio [5].

Since no mechanical work is done during shivering, virtually all the energy released by muscular contractions appears as heat. Nevertheless, shivering is not a very economical process from the thermoregulatory point of view because the blood flow to the superficial parts of the body is increased in the cold, which increases the chances for heat loss. Further, the constant movements increase convective heat loss. The efficiency of shivering, expressed in terms of the proportion of extra heat generated to that which is retained in the body, has been estimated as 48 percent [6]. However, the proportion of the heat which can be actually utilized to compensate the heat loss from the body is regarded as 10-11% only. Shivering is also fatiguing, like exercise, if prolonged.

7. Nervous control of shivering

Shivering is probably induced by changes in temperature both at the body periphery and in the body core, the slope of the metabolic increase being inversely related to these temperatures [7, 8]. The magnitude of the metabolic response to cold is proportional to the integral of the inputs from the periphery, core and CNS, i.e., a cool skin and warm body core will evoke less shivering than a cool skin and cool body core.

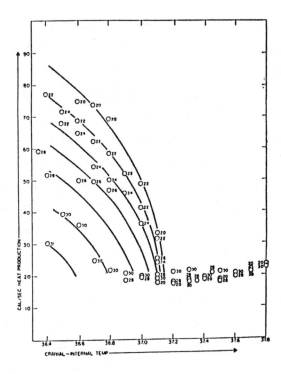

Fig. 3. Modification of the relationship between deep body temperature and metabolic rate due to shivering by different skin temperatures (from Benzinger [7]). (Numbers within the figure denote skin temperatures).

Figure 3 shows that the intensity of shivering depends on deep body temperature predominantly, while the threshold for shivering is modified by skin temperatures. It appears that hypothalamic temperature changes are about 10 times more effective in inducing heat production than temperature changes in the body periphery [9]. Intestinal cooling, resulting in a strong activation of the central input, is also very effective in inducing shivering.

Shivering in mammals and birds can also be elicited by direct cooling of the spinal cord [9]. The mutual relation of the spinal and peripheral variables is best expressed by a hyperbola. It follows from this relation that the lower the temperature of the skin, the higher the temperature of the spinal cord must be to induce shivering and vice versa.

Inputs from cold sensors are integrated in the preoptic-anterior hypothalamus. Impulses are transmitted from the caudal motor area to effector organs, such as the muscles (see chapter 6 for further details). Although shivering is involuntary and can occur in animals without a cerebral cortex, it can be inhibited by voluntary effort. This indicates that higher parts of the brain (e.g., from frontal sulcus) can also affect shivering. A suggested scheme for the nervous control of shivering is seen in Fig.4 (11). In addition to the primary motor center, located in the dorsomedial area of the hypothalamus, there exist secondary controlling centers which can initiate or inhibit shivering. The secondary centers are: (1) the cerebral cortex, which exerts an inhibitory effect; (2) an inhibitory center in the rostral hypothalamus (basal forebrain); (3) a facilitatory and inhibitory center in the septum; and (4) an inhibitory center in the ventromedial hypothalamus just ventral to the primary motor center.

From the caudal part of the hypothalamus, numerous connections course through the mid- and hindbrain, close to the rubrospinal or reticulospinal tracts down the ventrolateral columns of the spinal cord and to the muscles by the ventral roots.

The sympathetic nervous system may exert a fine control on shivering. Sympathetic ganglionic blocking agents have been reported to abolish shivering, while sympathetic activation may increase resistance to fatigue and increase the capacity for heat production by influencing the static discharge of cold receptors. Proprioceptive sensory input may also be enhanced, since appropriate stimulation of the sympathetic trunk increases the rate of discharge of spindle endings. Furthermore, norepinephrine affects the postsynaptic neuromuscular junction, inducing depolarization or hyperpolarization, increasing sensitivity of the skeletal muscle fiber membrane to acetylcholine and shifting the reverse potential of end-plate potentials.

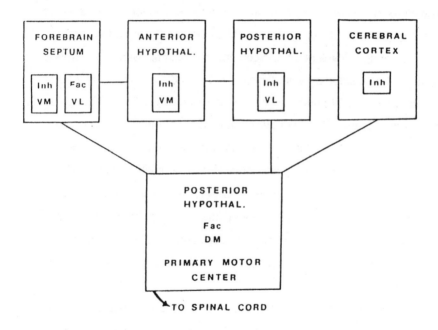

Fig. 4. Scheme of brain centers controlling shivering. Inh = inhibitory, Fac = facilitatory, VM = ventromedial, VL = ventrolateral, DM = dorsomedial area (from Hemingway[11]].

8. Summary

Shivering is a thermoeffector mechanism producing heat in cold-exposed homoiotherms to compensate for increased heat loss from the body.

Shivering represents an important defence response to cold and is a principal prerequisite of homeothermy - animals not exhibiting shivering may be considered primitive from the point of evolution.

Shivering is based on involuntary, repeated, and synchronous contractions of antagonistic groups of skeletal muscles (flexors and extensors) and is usually defined as an increase in reflex non-locomotor muscular tone with and without visible tremor.

Shivering is a nervous reflex and is induced both by central and peripheral thermal inputs. In mammals, the thermal inputs are integrated in the posterior hypothalamus and mediated by activation of somatomotor nerves.

Shivering occurs in all muscles of the body, except the facial, perineal, extraocular and middle ear muscles.

During muscle contractions, the energy of ATP is used as fuel. The amount of heat produced by shivering exceeds basal metabolic rate by a factor of 3 - 5. Larger species appear to produce more heat due to shivering.

Principal references:
1. Glossary of terms for thermal physiology. *Pflügers Arch* (1987) **410**: 567-587.
2. Kleinebeckel D and Klussmann FW (1990). Shivering. In: *Schönbaum E and Lomax P Thermoregulation: Physiology and Biochemistry*, Pergamon Press, New York, pp. 235-253.
3. Jung R, Doupe J, and Carmichael EA (1937). Shivering: a clinical study of the influence of sensation. *Brain* **60**: 20-38.
4. Stuart DG, Eldred E, Hemingway A, and Kawamura Y (1963). Neural regulation of the rhythm of shivering. In: *Temperature - Its Measurement and control in Science and Industry*, Vol. **3**, pp. 545 -557.
5. Himms-Hagen J (1976). Cellular thermogenesis. *Ann. Rev. Physiol.* **38**: 315-351.

6. Hardy JD (1961). Physiology of temperature regulation. *Physiol. Rev.* **41**: 521-606.

7. Benzinger TH (1969). Heat regulation: Homeostasis of central temperature in man. *Physiol. Rev.* **49**: 671 - 759.

8. Jessen C (1985). Thermal afferents in the control of body temperature. *Pharmac. Ther. Vol* **28**, pp. 107 - 134, 1985.

9. von Euler, C. (1961). Physiology and pharmacology of temperature regulation. *Pharmacol. Rev.* **13**: 361-398.

10. Simon E, Pierau FK and Taylor DCM: Central and peripheral thermal control of effectors in homeothermic temperature regulation. *Physiol. Rev.* **66.**, 235-300, 1986.

11. Hemingway A (1963). Shivering. *Physiol. Rev.* **43**: 397-422.

Self-evaluation questions for chapter 4.1.
1. What is shivering?
2. Where is shivering localized?
3. Can one record shivering?
4. What are the signals inducing shivering?
5. Where are the shivering control centers localized?
6. What is the metabolic efficiency of shivering?

Chapter 4.2

Learning Objectives
1. To understand the difference between obligatory, thermoregulatory and diet-induced nonshivering thermogenesis.
2. To realize that nonshivering thermogenesis is localized to brown adipose issue.
3. To understand the acute effect of stimulation of brown adipose tissue.
4. To understand the function of the uncoupling protein.
5. To realize the existence of the recruitment process of nonshivering thermogenesis.
6. To discuss the significance of nonshivering thermogenesis in newborn infants and adult man.
7. To vizualize clinical and therapeutical consequences of alterations in the capacity for nonshivering thermogenesis.

Bullets

B-1   Thermoregulatory nonshivering thermogenesis is located in brown adipose tissue.
B-2   Capacity for nonshivering thermogenesis can be measured as the metabolic response to norepinephrine injection.
B-3   Brown adipose tissue is widely distributed in the body.
B-4   Brown adipose tissue is sympathetically innervated.
B-5   Thermogenesis is stimulated via β_3-receptors and an increase in cellular cAMP levels.
B-6   The uncoupling protein permits re-entry of H^+ into the mitochondrial matrix.
B-7   The capacity for nonshivering thermogenesis is determined by the total amount of UCP.
B-8   Cold acclimation leads to the recruitment of brown adipose tissue, especially to more UCP.
B-9   Precocial neonates are born with a high capacity for nonshivering thermogenesis.
B-10 Altricial neonates acquire an increased capacity for nonshivering thermogenesis because they are exposed to the cold.

B-11 Adult man has brown adipose tissue.
B-12 The functional significance of brown adipose tissue in adult man is not established.
B-13 Less brown fat gives less cold tolerance and a higher propensity for obesity.
B-14 Activation of nonshivering thermogenesis in brown adipose tissue may be a treatment of obesity.

Chapter 4.2

NONSHIVERING THERMOGENESIS AND BROWN ADIPOSE TISSUE

BARBARA CANNON AND JAN NEDERGAARD
University of Stockholm, Sweden

The advantages of having a high and constant body temperature have been elucidated in chapters 1 and 2. In endothermic homeotherms, i.e., in most mammals under all normal conditions (the only exception really being mammals that enter hibernation or torpor), core temperature is defended in the cold by a combination of behavioural defense strategies, limitation of heat loss through vasoconstriction and piloerection and, when necessary, an activation of heat production. This thermoregulatory heat production derives either from shivering thermogenic processes (as outlined in chapter 4.1) or from nonshivering thermogenic processes, to be discussed here.

Mammals are basically energy-conserving and therefore – with increasing cold – initially utilise behavioural mechanisms and vasoconstriction before they activate the energy-consuming thermogenic processes. They are also comfort-seeking and therefore – although unconsciously – activate their capacity for nonshivering thermogenesis (which proceeds unnoticed, probably also by the mammals themselves) prior to activating shivering (which is much more disturbing for other activities). In the present chapter, the mechanism of nonshivering thermogenesis will be described and its regulation discussed.

1. Nonshivering Thermogenesis

All processes of combustion in the body result in heat production and can therefore quite correctly be called thermogenic. Indeed, basal metabolism takes place without much muscle movement and can therefore be considered a form of nonshivering thermogenesis, sometimes referred to as obligatory nonshivering thermogenesis. However, in the present context, we are not referring to such processes but only to those processes which are initiated specifically in order to maintain body temperature and that can thus be called *thermoregulatory nonshivering thermogenesis*. This is thus the heat produced above basal metabolism, in the absence of muscle activity and with the purpose of producing heat.

Although not intrinsically self-evident, it has turned out that another homeostatic mechanism, *diet-induced nonshivering thermogenesis (DIT)*, is mechanistically very similar to thermoregulatory nonshivering thermogenesis. Whereas in thermoregulatory nonshivering thermogenesis energy is used to produce heat in order to defend the body temperature "set-point", in DIT energy is utilized apparently with the purpose of defending the body energy "set point" (i.e. body weight), and the heat is a waste product.

However, these two processes share the same effector mechanism: sympathetic activation of brown adipose tissue.

1.1 Determination of the capacity of nonshivering thermogenesis

It was shown more than thirty years ago that the thermoregulatory increase in heat production is mediated by norepinephrine released by the sympathetic nervous system. The capacity of an animal to demonstrate nonshivering thermogenesis can thus be determined by the injection of norepinephrine at its thermoneutral temperature. The resulting increase in heat production (or oxygen consumption) is then defined as the nonshivering thermogenic capacity of that animal.

At least in experimental animals, the extra heat production elicited by norepinephrine injection, i.e., the thermoregulatory nonshivering thermogenesis discussed here, is fully the consequence of brown adipose tissue activation: most of the heat production occurs there, and the remainder comes from the secondary activation of heart and lung muscles. Hence, animals without brown adipose tissue are incapable of this type of thermoregulatory nonshivering thermogenesis. It may be so that birds (that fully lack brown adipose tissue) have developed an alternative mechanism, but it is still not even fully established that birds posses thermoregulatory nonshivering thermogenesis.

If an animal has been living at its thermoneutral temperature for a prolonged period in a long light cycle and is not overfed – and thus theoretically has no requirement for thermoregulatory nonshivering thermogenesis – norepinephrine nonetheless elicits a small increase in heat production. It is currently unclear if this heat production conforms with that described above as arising from brown adipose tissue. Under all other conditions, the heat production is directly proportional to the amount of active brown adipose tissue in the animal.

2. Brown Adipose Tissue

The Swiss naturalist Konrad Gessner was probably the first to describe (in 1551) what must have been brown adipose tissue, a tissue which is "neither flesh nor fat" in the winter-acclimated marmot. The role of the tissue remained equivocal until the outstanding work of Robert E. Smith in the early 1960's, in which it was clearly shown that the function of the tissue was to generate heat.

Brown adipose tissue has been found in all mammalian neonates, with the exception of the domesticated pig. It is present also in cold-acclimated adult rodents and probably has the potential to regenerate in adulthood in all species in which it was present in the perinatal period. The tissue is dispersed in a number of depots, such as those indicated in Fig. 1. The majority are located in the thorax, such that the heat produced can readily warm vital organs, but an important depot is also located perirenally. It is also now generally accepted that many depots of typical white adipose tissue may be infiltrated with small quantities of brown adipose tissue and possess brown adipocyte stem or precursor cells.

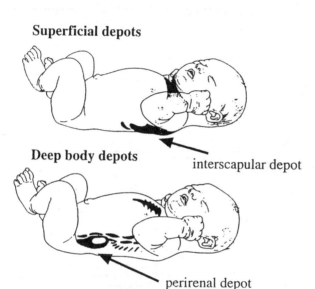

Fig. 1. Depots of brown adipose tissue found in the newborn human infant. Similar depots are found in most experimental animals. For experimental studies in animals, the interscapular depot is the most easily accessible. The drawing is modified from Dawkins and Hull (1965).

Depending on the degree of activation and the triglyceride content, the colour of "brown" adipose tissue varies between buff-beige and reddish-brown. The tissue is highly vascularised, indicating a high requirement for oxygen delivery to the cells, and the tissue colour, in part, derives from the high blood content. When the tissue is at its most active, the blood flow is among the highest known for any organ. Vasodilation seems to be regulated by a metabolite of the cells themselves, and both nitric oxide and adenosine have been suggested in this context. The high blood flow also permits transfer of the heat from its site of formation, much like a central heating system.

As stated above, even before the function of brown adipose tissue was deduced, it was found that the agent responsible for initiating nonshivering thermogenesis was norepinephrine, the sympathetic nervous system transmitter. In accordance with this, brown adipose tissue has a dense sympathetic innervation, of at least two types (Fig. 2). The blood vessels are innervated with sympathetic fibres containing both norepinephrine and NPY; these fibres are presumably involved in the maintenance of vascular tone. Another system of sympathetic fibres innervates all the individual adipocytes with boutons en passant. These fibres are those involved in the direct regulation of tissue activity (heat production).

The sensory transmitter substance P and also CGRP (calcitonin gene related peptide) are found in the tissue but their relationship to the two fibre systems described is not clear. The tissue appears not to possess parasympathetic innervation.

The principal organisation of the innervation is indicated in Fig. 2. The sympathetic fibres in the tissue derive from the (cervical) sympathetic ganglia. It is evident that several hypothalamic centres can influence the sympathetic activity of the tissue.

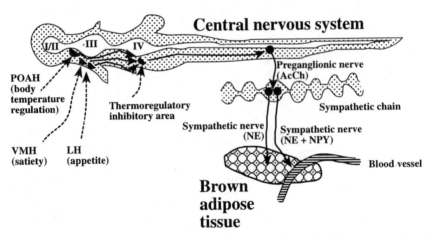

Fig. 2. The principal organization of the innervation of brown adipose tissue. Both the preoptic anterior hypothalamic (POAH) temperature control centers and the ventromedial and lateral hypothalamic (VMH, LH) food intake centers influence brown adipose tissue activity. The sympathetic nerves to the tissue contain norepinephrine (NE) and those to the blood vessel also neuropeptide Y (NPY).

The brown adipocytes themselves contain multiple small fat droplets (and are therefore said to be multilocular adipocytes in contrast to the unilocular white adipocytes), and have a centrally located nucleus. The remainder of the cytoplasm is notably filled with mitochondria having very dense cristae (Fig. 3). The high mitochondrial density is the second reason for the brownish colour of the tissue. The dense cristae are unusually well-endowed with respiratory chain enzymes which can permit high rates of substrate oxidation.

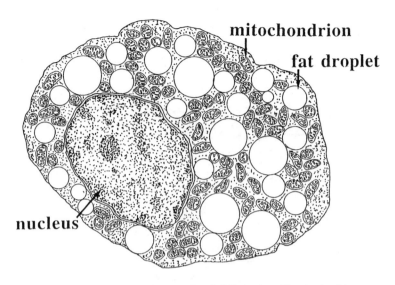

Fig. 3. A brown fat cell. Note how the cytoplasm is fully dominated by mitochondria.

3. Mechanism of Acute Heat Production

3.1 Cytosolic events

When a decrease in core temperature is perceived by the hypothalamus (see Chapter 6), the sympathetic innervation to brown adipose tissue is activated. This results in the release of norepinephrine in areas adjacent to the brown adipocytes. The norepinephrine may bind to different adrenergic receptors. However, for the activation of heat production, the binding to a class of β-adrenergic receptors, known as $β_3$-receptors, is the most important. These receptors couple to adenylate cyclase (Fig. 4), and the second messenger, cyclic-AMP, is produced. Through activation of protein kinase A, the enzyme hormone-sensitive lipase is phosphorylated, which results in its activation. The lipase degrades the endogenous triglycerides of the fat droplets, releasing free fatty acids, which serve as substrates for mitochondrial oxidation (Fig. 4).

3.2 Mitochondrial events

In all mitochondria, substrate oxidation results in the production of a proton gradient across the mitochondrial inner membrane, and it is in this form that the chemical energy from the substrate is temporarily stored (Fig. 5A).

In mitochondria in general, substrate oxidation only proceeds if the cell has utilised its reserves of ATP and has therefore generated ADP. The presence of ADP in the mitochondrial matrix results in the re-entry of protons into the matrix through the enzyme ATP synthase and the formation of ATP (Fig. 5A). The need to regenerate ATP therefore drives substrate oxidation (=respiration), but only until all the ADP is converted to ATP. In such a system, heat is only produced if ADP is continually regenerated by an ATP-utilising process. This is not the mechanism found in brown-adipose-tissue mitochondria.

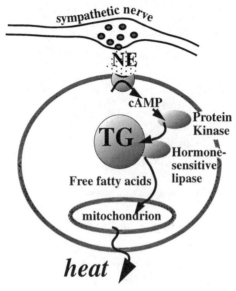

Fig. 4. The acute effect of norepinephrine (NE) on the brown-fat cell. TG: the triglycerides constituting the fat droplets.

An alternative means of heat production, which circumvents the need for an ATP-utilising reaction, is to simply short-circuit the proton circuit described above. Under such circumstances, the protons pumped out of the mitochondrial matrix by oxidation are allowed to re-enter the matrix and, in so doing, dissipate the stored energy in the form of heat. This is the mechanism utilized in brown-adipose-tissue mitochondria. Inserted into the inner membrane of brown adipose tissue mitochondria is the specialised proton-translocating protein, thermogenin or uncoupling protein (UCP), the function of which is indeed to dissipate the proton gradient and thus permit high rates of respiration = heat production (Fig. 5B).

The UCP gene is expressed only in brown adipose tissue. It belongs, however, to a large family of mitochondrial inner membrane transporter proteins, among them, e.g., the oxoglutarate carrier and the adenine nucleotide translocase. There exist also proteins with no presently demonstrated physiological function that are more similar to UCP than to any other protein. These proteins are referred to as UCP2, UCP3, etc.

(UCP becomes UCP1 in this terminology). Members of this UCP-like family are also found in other tissues than brown adipose tissue and even in plants.

UCP has a molecular weight of approximately 32 kD. The amino acid sequence is known in several species and is highly conserved. Still not completely understood at the molecular level, however, is the acute control of UCP activity, in spite of the fact that this activity has been experimentally reconstituted in liposomes.

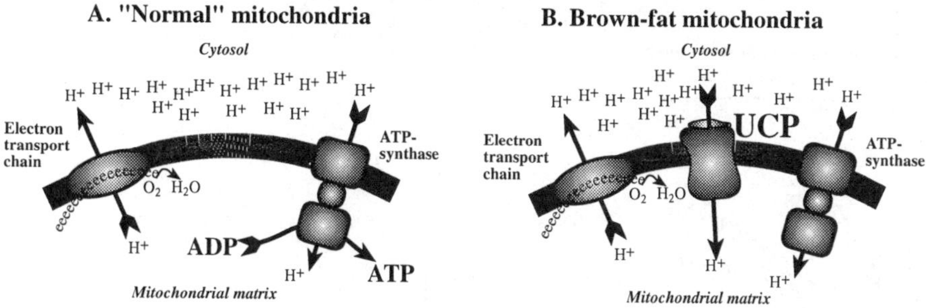

Fig. 5. Mitochondrial function. In "normal" mitochondria (A), the H^+ gradient generated by the respiratory system, is utilized primarily for the production of ATP from ADP. In brown-fat mitochondria (B), the uncoupling protein (thermogenin, UCP) allows (when activated) for a functional short-circuit of protons through the mitochondrial membrane, releasing the energy stored in the H^+ gradient as heat. The exact way in which UCP transports H^+, is not known; a possibility is that UCP is really a fatty acid transporter. According to this view, a fatty acid, together with a H^+, enters from the cytosol spontaneously directly through the mitochondrial membrane, and it is the retransport of the fatty acid (in the anion form) back to the cytosol that is catalysed by UCP; the net action of UCP is still in this view to transport H^+, but the fatty acid acts as a shuttle.

4. Regulation of the Total Capacity for Heat Production

As indicated above, an animal's capacity for nonshivering thermogenesis is directly proportional to the amount of active brown adipose tissue it possesses; more accurately, it is directly proportional to the total amount of UCP in the animal.

When an animal encounters a cold environment in which it must elevate heat production in order to defend its body temperature, it will first activate heat production in its already existing brown adipose tissue. If this capacity is insufficient to fully counteract heat loss, shivering will also be initiated (Figs. 6A and B). However, simultaneously with this, a series of reactions will also be initiated in the brown adipose tissue, the goal of which is to increase the capacity for nonshivering thermogenesis. This increased (or recruited) capacity will thus with time alleviate the need for shivering, – and also provide an increased cold tolerance, should the environmental temperature decrease further.

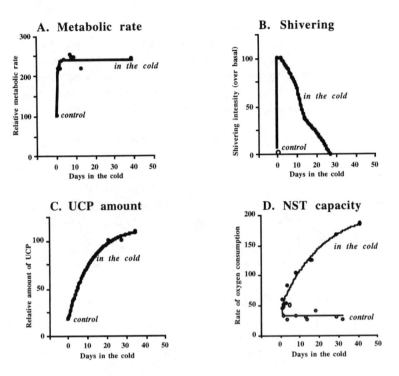

Fig. 6. Recruitment of brown adipose tissue and of nonshivering thermogenesis during acclimation to cold. A: Immediately when a mammal (here a rat) is exposed to a cold environment, its metabolic rate is increased and this rate remains high during the cold exposure. B: Although the immediate increase in metabolism is due to the increase in shivering intensity, shivering gradually wanes, but metabolism remains high (i.e., nonshivering thermogenesis). C: During the same time, the total content of UCP in the brown adipose tissue of the mammal increases. D. The increase in UCP endows the animal with an increased capacity for nonshivering thermogenesis. Detailed references in Nedergaard et al. (1995).

This recruitment process entails increases in cell number, mitochondrial content, and total content of UCP. In parallel, to support the enhanced thermogenic activity, there occur increased innervation and vascularisation. In Fig. 6, the changes in various parameters with time are indicated. The elevated metabolism necessary to defend body temperature remains unchanged throughout a prolonged period in the cold, while shivering successively decreases and is replaced by an increased nonshivering thermogenesis, the capacity of which can be estimated in the animal both from its UCP content and from its metabolic response to norepinephrine injected in a thermoneutral environment.

A physiologically meaningful but biologically unexpected conclusion is that it is the norepinephrine released from the sympathetic nerves that also induces this recruitment process.

Norepinephrine thus potentiates cell proliferation in dormant precursor cells in the tissue; it may also decrease the activity of apoptotic processes. Cell proliferation in the precursor cells is mediated through activation of β_1-adrenergic receptors, while a decreased cell death rate is a β_3-adrenergically-mediated process in more mature cells. Also promotion of mitochondriogenesis and cell differentiation in general are stimulated by norepinephrine. The activation of the expression of the UCP gene is mainly β_3 adrenergically mediated, although the presence of thyroid hormone is also necessary. In the control of both cell proliferation and gene expression, the numerous α_1-adrenergic receptors may also play an important auxilliary role.

5. Brown Adipose Tissue in Neonates

All mammalian neonates are in a much more perilous situation with respect to body temperature regulation than the corresponding adults of the same species. They are often poorly insulated, both above and below the skin (i.e., they are hairless and lack subcutaneous fat), and they are also initially wet. Vasoconstrictor mechanisms are poorly developed, and the body mass is small compared with the large surface area, i.e., the ratio between basal metabolic heat produced and heat loss from the skin is low. It is consequently not surprising that, in mammals, brown adipose tissue is normally most active during this age period.

It is possible to categorise mammalian neonates into three groups, depending on their degree of development at birth. It has turned out that the ability of neonates to defend body temperature follows a similar division, and that this is primarily related to the degree of development of their brown adipose tissue (Fig. 7).

The first group, the *precocial neonates*, e.g., lambs, guinea pigs, consists of rather well-developed animals at birth. They have fur, their eyes are open, they can walk, and often have few or no siblings. Already at birth, they can elevate metabolism in response to a cool environment and show a clear norepinephrine-induced increase in metabolism, indicating well-developed brown adipose tissue. It follows that the total content of UCP is highest at the time of birth. This is in itself something of an enigma as it is presently not clear what the physiological signal is that stimulates UCP gene expression in utero. The capacity for nonshivering thermogenesis decreases as the animals grow.

The second group, the *altricial* ("nest-dependent") *neonates*, consists of less-developed animals than the precocial ones; rats and mice belong to this category. There are often many siblings, they are naked at birth, their eyes are closed, and they can barely move. They do elevate metabolism in a cool environment, but only to a small extent, and are rapidly overwhelmed by the cold. Norepinephrine elicits a small increase in metabolism in a thermoneutral environment. The responses to cold and to norepinephrine increase markedly in the days after birth, and the total content of UCP increases in parallel. These responses to a cool environment closely resemble those of adults and are presumably mediated in the same way, i.e., through sympathetic activation as a response to the cold.

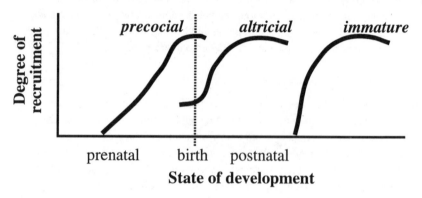

Fig. 7. The perinatal development of brown adipose tissue and thus of the capacity for non-shivering thermogenesis in three different categories of mammals.

The third group, the *immature neonates*, e.g., hamsters, show no enhanced metabolism in a cool environment, nor do they respond to a norepinephrine injection. It is not possible to detect any UCP in the adipose tissue depots of these animals. Detectable levels of UCP are observed about 10 days postnatally, and at this time, temperature regulation responses can also be measured.

It has been difficult to assign *the human infant* to any of these groups, as there are only a few reports in the literature that can be used to clarify the question. Human infants are clearly no exception to the rule that brown adipose tissue is of greatest importance in newborns. Infants do show an elevated metabolism in a cool environment, as well as a higher skin temperature over the region of the interscapular brown adipose tissue pad. UCP is present in this tissue, even before birth, but probably increases after birth. In a unique study, Karlberg and collaborators (1965) showed that a 4-day old infant had a somewhat higher response to infused norepinephrine than the same infant on day 1. Taken together, these results indicate that the human infant is probably altricial in nature, i.e., he shows an increasing thermogenic capacity in the postnatal period, as an adaptive response to the relative cold to which he is exposed.

6. Brown Adipose Tissue in Adult Humans

It is undoubtedly so that human infants have an active brown adipose tissue which, in all respects, resembles that found in smaller mammals. It is also evident that, in smaller mammals, brown adipose tissue and nonshivering thermogenic capacity decline as the animal increases in size and insulation. However, should such a small mammal be once again exposed to an environment which it experiences as cold, its brown adipose tissue becomes recruited, and the capacity for nonshivering thermogenesis once again increases. The annual temperature cycle at temperate latitudes causes this cycling pattern in thermogenic capacity throughout life, albeit this capacity seems to decline somewhat in old age. It would, therefore, be most reasonable to propose that a similar potential to recruit brown adipose tissue should also follow a human adult throughout life, but that due to our larger size, and our more favourable surface-to-volume ratio, this potential is rarely utilized.

In an attempt to ascertain whether adult humans indeed demonstrate a capacity for nonshivering thermogenesis, certain criteria may be enumerated which, based on the observations in rodents, would have to be fulfilled. These are simply the presence of brown adipose tissue (which in reality is synonymous with the presence of UCP) and that acclimation to cold should lead (Fig. 6) to recruitment, i.e., decreased shivering, increased response to norepinephrine, and increased amount of UCP.

6.1 The presence of Brown adipose tissue and UCP

There are a number of reports indicating the presence of brown-adipose-tissue-like tissues in adult humans of all ages. It has also been suggested that the amount is greater in cold-exposed individuals. However, these results as yet have only been obtained at the light microscopic level by determination of fat cells showing high multilocularity. Since white fat cells under certain conditions may also show multilocularity, these reports can not be taken as evidence of active brown adipose tissue. Neither, however, do they preclude this possibility.

As noted above, UCP is certainly present in infants. It has also been verified immunologically that UCP occurs in adult humans under certain pathological conditions of high sympathetic drive (see below), and the presence of UCP in normal or cold-exposed adults has also been reported, both immunologically and at the mRNA level. However, whether sufficient UCP is present in normal adults to have a significant effect on metabolism or cold tolerance is not easy to determine.

The results from the pathological conditions demonstrate, however, unequivocally that humans can recruit brown adipose tissue again in adult life, provided the stimulus is sufficiently strong. They do not, however, demonstrate a physiological role for the tissue in cold acclimation.

6.2 Recruitment effects

There are a few studies which do seem to imply a recruitment of nonshivering thermogenesis due to cold acclimation in humans. Thus, it has been reported in humans that shivering in response to acute cold is decreased in winter compared with summer, but since no measurements of metabolism were made, this report remains inconclusive. However, in another series of studies, daily cold exposure for 8 hours resulted in gradually decreasing shivering with time, but a maintained elevated metabolism (Fig. 8). This decreased shivering with maintained elevated metabolism strongly resembles the results from cold-acclimated rodents (Fig. 6). Also, apparent nonshivering thermogenesis in humans has been demonstrated to be particularly sensitive to hypoxia, a feature which has also been observed in animals. There is even a report of an elevated response to norepinephrine in cold-acclimated adults compared to controls; as already described, this is accepted in rodents as indicating the presence of potentially active brown adipose tissue. No UCP measurements, however, have ever been reported, and they are, of course, not so easy to obtain.

Nonetheless, taken together, these reports suggest that adult humans exposed to chronic cold, with its concomitant elevation of sympathetic activity, probably do recruit active brown adipose tissue and thus thermoregulatory nonshivering thermogenesis, which can contribute to body temperature regulation.

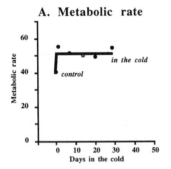 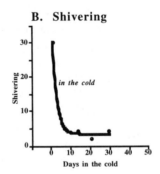

Fig. 8. Effect of acclimation to cold in humans: possible induction of thermoregulatory nonshivering thermogenesis. When humans are exposed to cold, they increase their metabolism (A) (although less than, e.g., a rat, due to their better surface/volume ratio). The increased metabolism persists even though shivering decreases (B). Adapted from Davis (1961).

7. Pathological Conditions

7.1 Conditions with more Brown adipose tissue

As noted above, there are reports of the occurrence of confirmed brown adipose tissue in adult humans under certain pathological conditions. There is a rare and benign tumour, called a hibernoma, of unknown etiology, in which the tissue mass is clearly brown adipose tissue. The rare adrenal tumour phaeochromocytoma, which results in massively elevated circulating catecholamine levels, also leads to the recruitment of confirmed depots of brown adipose tissue. It may be presumed that these catecholamine levels mimic those released locally in the tissue on sympathetic nerve activity following, e.g., severe cold stress. The trypanosome-mediated Chargas disease also causes sympathetic overactivity and has been reported to increase brown adipose tissue. Certain other pathological conditions characterised by elevated catecholamine levels or sympathetic hyperactivity have also been reported to correlate with the confirmed presence of UCP in adipose tissue depots, thus defining them as brown adipose tissue. Whether these conditions are connected with an increased capacity for nonshivering thermogenesis has not been examined.

7.2 Less Brown adipose tissue: Obesity

A low amount of brown adipose tissue should lead to a low capacity, not only for thermoregulatory nonshivering thermogenesis, but also for diet-induced nonshivering thermogenesis (as both of these processes are located to brown adipose tissue). No such conditions have been demonstrated in humans, but in rodents, all genetically obese strains also have atrophied brown adipose tissue and a decreased capacity for thermoregulatory and diet-induced thermogenesis. This lowered capacity for nonshivering thermogenesis contributes to the obesity of these genetically obese animals (e.g., the *ob/ob* and *db/db* mouse and the *fa/fa* rat), as apparently less of the food consumed is converted to heat. Whether or not such a relationship exists in humans – i.e., whether some humans have a low capacity for nonshivering thermogenesis and, therefore, become obese on the same caloric intake that is adequate for others – is currently a subject of considerable speculation.

7.3 More Brown adipose tissue as obesity treatment

However, from the point of view of obesity treatment, the important point is that adult humans can recruit brown adipose tissue, not whether or not their obesity may originally result from inactivity of the tissue. Thus, a pharmacological regimen which would mimic the effects of severe chronic cold or massively elevated catecholamine levels, without their attendant side-effects, might be expected to recruit and activate brown adipose tissue and, hence, lead to an increased nonshivering thermogenesis and increased energy consumption. If this increase in energy consumption were not compensated by an increase in energy intake (not an elementary assumption, as animals exhibiting thermoregulatory nonshivering thermogenesis do compensate for the energy loss by a higher food intake), this could lead to body mass reduction. It is because of this potential therapeutic effect that so much attention is being directed lately to the study of the subtype of β-adrenergic receptors responsible for stimulation of nonshivering thermogenesis. There is still not a consensus that the receptors regulating thermogenesis in humans belong to the $β_3$-subtype. Other possibilities, such as other ways of increasing the expression of UCP, or activating transcription factors or coactivators involved in (brown) adipose tissue differentiation (e.g., PPARγ and PGC-1) are therefore also actively being pursued at the moment.

B-14

Despite these difficulties, there is clearly a future for brown adipose tissue in a context which is an extension of, and in a way peripheral to, the role of the tissue in thermoregulatory nonshivering thermogenesis and body temperature regulation. However, as a necessary side-effect of any such recruitment-stimulating treatment, the subjects will also exhibit an increased cold tolerance.

8. Suggested Reading

Cannon B, Nedergaard J (1985). The biochemistry of an inefficient tissue: brown adipose tissue. *Essays Biochem.* **20**:110-164.

Foster DO, Frydman ML (1978). Nonshivering thermogenesis in the rat. II. Measurements of blood flow with microspheres point to brown adipose tissue as the dominant site of the calorigenesis induced by noradrenaline. *Can. J. Physiol. Pharmacol.* **56**:110-122.

Himms-Hagen J (1990). Brown adipose tissue thermogenesis: role in thermoregulation, energy regulation and obesity. In: Schönbaum E, Lomax P (eds), *Thermoregulation: Physiology and Biochemistry*. Pergamon Press, New York, pp. 327-414.

Jansky L (1973). Nonshivering thermogenesis and its thermoregulatory significance. *Biol. Rev.* **48**:85-132.

Klaus S (1997). Functional differentiation of white and brown adipocytes. *Bioessays* **19**:215-223.

Lean MEJ, James WPT (1986). Brown adipose tissue in man. In: Trayhurn P, Nicholls DG (eds), *Brown Adipose Tissue*. Edward Arnold, London, pp. 339-365.

Nedergaard J, Lindberg O (1982). The brown fat cell. *Int. Rev. Cytol.* **74**:187-286.

Nedergaard J, Connolly E, Cannon B (1986). Brown adipose tissue in the mammalian neonate. In: Trayhurn P, Nicholls DG (eds), *Brown Adipose Tissue*. Edward Arnold Ltd., London, pp. 152-213.

Nedergaard J, Cannon B (1992). The uncoupling protein thermogenin and mitochondrial thermogenesis. In: Ernster L (ed), New Comprehensive Biochemistry vol. 23: Molecular Mechanisms in Bioenergetics. Elsevier, Amsterdam, pp. 385-420.

Nedergaard J, Herron D, Jacobsson A, Rehnmark S, Cannon B (1995). Norepinephrine as a morphogen? - its unique interaction with brown adipose tissue. *Int. J. Dev. Biol.* **39**:827-837.

Ricquier D, Bouillaud F (1997). The mitochondrial uncoupling protein: structural and genetic studies. *Prog. Nucleic Acid Res. Mol. Biol.* **56**:83-108.

Smith RE, Horwitz BA (1969). Brown fat and thermogenesis. *Physiol. Rev.* **49**:330-425.

Self-Evaluation Questions

1. What are the differences and the similarities between thermoregulatory and diet-induced nonshivering thermogenesis?
2. How is the capacity for nonshivering thermogenesis determined?
3. Why is brown adipose tissue brown?
4. What is the difference between the sympathetic nerves innervating the brown fat cells themselves and the blood vessels of the tissue?
5. What restricts the rate of substrate utilization in normal mitochondria?
6. What does the abbreviation UCP mean?
7. What is the recruitment process?
8. Compare precocial, altricial and immature newborns with respect to thermoregulation.
9. What criteria can be set for the establishment of a role of nonshivering thermogenesis in cold acclimation of adult man?
10. What are the metabolic consequences of a reduced amount of brown adipose tissue?

Chapter 5

Learning Objectives
1. To understand the physical and environmental factors which modify heat loss.
2. To understand the physiological regulation of heat loss, separating autonomic and behavioral regulation.
3. To understand evaporative heat loss: insensible perspiration, sweat secretion and mechanisms.
4. To understand the regulatory mechanism of cutaneous blood flow.

Bullets
B-1   Heat loss depends on evaporative and non-evaporative mechanisms.
B-2   About 20% of heat produced under resting conditions is lost by insensible perspiration.
B-3   Heat loss from the skin by sweating is the main route of heat loss in high environmental temperature.
B-4   Thermal sweating is mainly controlled by core temperature through cholinergic sympathetic nerve fibers.
B-5   NaCl is lost from sweat. But a fit person has a high sweat rate and a low sweat NaCl concentration.
B-6   In a thermoneutral windless environment, about 60% of heat loss is by radiation.
B-7   The low thermal conductivity of air is important for thermal insulation.
B-8   Forced convection of air is effective to increase heat loss.
B-9   Skin blood flow is controlled by sympathetic vasoconstrictor nerve in the distal limbs.
B-10   The arterio-venous anastomoses and countercurrent heat exchange play important roles in heat exchange in the distal limbs.
B-11   Thermoregulation competes with homeostatic mechanisms for maintaining circulation and body fluid balance.

Chapter 5

HEAT LOSS MECHANISMS

TAKETOSHI MORIMOTO

Department of Physiology, Kyoto Prefectural University of Medicine
Kamigyoku, Kyoto 602, Japan

1. Introduction

Body temperature is maintained at a constant level when the internal and/or external heat loads equal the heat loss to the environment. Heat exchange between the human body and its environment is attained by evaporative heat loss and non-evaporative heat exchange. Non-evaporative heat loss is the sum of heat flow or flux due to radiation, convection and conduction from a body to the environment, and it is called either dry, Newtonian or sensible heat loss (see Chap. 2). Up to the upper critical temperature, the required level of non-evaporative heat loss in human is regulated by modifying skin temperature through alteration of the cutaneous blood flow. At an environmental temperature higher than the upper critical temperature, evaporative heat loss becomes the main avenue of heat loss. Table 5.1 outlines the mode and heat loss partitioning at various ambient temperatures for specific environmental conditions along with the mechanisms -autonomic and/or behavioral - of thermoregulation mediating them. The relative effectiveness of these avenues of heat loss changes with variations in environmental conditions. As shown in the table, animals, including humans, control these pathways by autonomic regulation and behavior to maintain body temperature at a required level.

In this chapter, these heat loss mechanisms will be outlined and discussed with reference to the evaporative and non-evaporative heat exchange channels.

2. Evaporative Heat Loss

As the environmental temperature rises and exceeds the upper critical temperature, non-evaporative heat loss is progressively reduced, and the dependency upon evaporative heat loss increases. Water has a high heat of vaporization, and it is the only mechanism available for the reduction of body temperature when the environmental temperature is higher than that of the body surface. The quantity of heat absorbed by water during the process of vaporization, namely the latent heat of vaporization for water, is given by the equation:

$$\lambda = 2490.9 - 2.34T \; [J \cdot g^{-1}] \tag{1}$$

where T is temperature of water in °C. Thus, evaporation of 1 g of water from the body surface at a temperature of 30 °C absorbs 2.43 kJ (580 cal·g^{-1}) of heat.

Table 5.1 Heat loss and its modification

Environmental factors	Mode of heat transfer	Partition of heat loss* 25°C	30°C	35°C	Autonomic regulation	Behavioral regulation
Temperature *Sunshine*	Radiation	67%	41%	4%	skin blood flow decrease (L) increase (H)	sun (L) or shade (H) seeking use of heater (L) change in radiating surface by posture (L, H)
Temperature *Air movement*	Conduction + Convection	10%	33%	6%	skin blood flow (L,H) piloerection (L)	clothing & housing (L) air movement (H) air conditioning (L, H) huddling, nesting (L)
Humidity *Air movement* *Temperature*	Evaporation	23%	26%	90%	sweating (H) panting (H)	skin wetting by saliva spreading, wallowing, etc. (H)

*: the values are approximations for a nude human subject in an environment with a constant, low air movement. L and H between parentheses indicate response to low or high temperature, respectively.

Depending on the species, evaporative heat loss takes place from either the skin or the mucosal surface of the upper respiratory tract, leading to cooling of the blood flowing in the immediate proximity. In humans, the water that evaporates from the skin comes from insensible perspiration and autonomic sweat secretion, while in some other animals it is derived behaviorally, such as by saliva spreading for rats, wallowing for pigs, and sprays for elephants.

Insensible perspiration, the passive evaporation from the surface of the body of water that has diffused through the skin (as opposed to water excretion in sweat), leaves the human body at the rate of 900 g·day^{-1} in temperate conditions, and includes diffusive water loss from the skin (about 500 g·day^{-1}) plus weight lost as water vapor from the lung (about 400 g·day^{-1}). Insensible perspiration plays an appreciable role in heat dissipation: 500 g x 2.43 kJ = 1215 kJ·day^{-1} (290 kcal·day^{-1}) from the skin, and 400 g x 2.43 kJ = 972 kJ·day^{-1} (232 kcal·day^{-1}) through respiration under sedentary conditions in thermoneutral conditions. Overall, it accounts for about 20% of the body's resting metabolism. Under conditions of high environmental temperatures and during exercise, the chief means of evaporative heat loss from the skin takes place via the evaporation of water in sweat. Man can sweat copiously, which enables control of body temperature and maintenance of body functions in hot environments. Subjects well acclimatized to heat by repeated exposure to heat (see Chap. 11.1) can sweat at a rate of more than 2,000 ml·hr^{-1} for limited periods of time. This is equivalent to the heat loss of 4,860 kJ (1,161 kcal) when all the sweat is evaporated from the skin. However, the rate of heat loss by evaporation (He) is dependent upon the gradient between the water vapor pressure of the skin (Psk) and that of the environment (Pa), the percentage of skin surface wetted with sweat (Aw), and a factor (he) determined by other factors, including air velocity, as:

B-2

$$He = he \cdot (Psk - Pa) \cdot Aw. \qquad (2)$$

B-3

Thus, heat is dissipated by evaporation when the water vapor pressure at the skin surface is greater than that of the environment.

2.1. Sweat glands and sweating mechanism

Thermoregulatory sweat in humans is secreted by eccrine sweat (atrichial) glands that cover the hairless area of the body surface. The number of sweat glands in humans ranges from 2 to 5 million; it is higher in individuals raised in tropical areas up to 2 years of age. The density is greatest on the sole, palm and forehead skin, and least on the trunk. However, the secretory activity does not parallel sweat gland density. Larger sweat glands can secrete sweat more copiously than smaller ones. The sweat secretion for a given region of skin is thus dependent upon both the density of sweat glands and sweat secretion per gland, with the chest and back often exhibiting the highest sweat rates.

Eccrine sweat glands consist of a secretory coil and an excretory duct which conveys sweat to the skin surface through the dermis. The secretory coils are composed of clear, dark mucoid, and myoepitelial cells. The clear cells have well-developed basal membrane infoldings with abundant mitochondria, which indicates that the clear cells are involved in the secretion of isotonic precursor sweat. The function of the dark cells is unknown. It is thought that the myoepithelial cells provide mechanical support for the secretory coil wall against the increase in laminar hydrostatic pressure during sweating, expelling precursor sweat to the excretory duct. The sweat ducts consist of two layers of cells, the basal and laminar cells. The basal ductal cells are replete with mitochondria, and the cell membrane is covered with Na^+ pumps for ductal Na^+ absorption. It has been suggested that the laminar ductal cells, which have a dense layer of tonofilaments, provide the structure of the tubular lumen. Thus, the precursor sweat produced in the secretory coils is isotonic with the plasma, and during the passage to the skin surface through the duct, NaCl is reabsorbed, yielding hypotonic sweat. This process is facilitated by aldosterone.

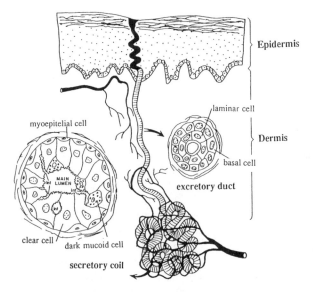

Figure 5.1 Diagram of eccrine sweat gland unit, the secretory coil, and excretory duct.

Isotonic precursor sweat is derived from interstitial fluid and blood and secreted by clear cells into intercellular canaliculi (Int) and the lumen of the secretory coil. During passage of the precursor sweat through the excretory duct to the skin surface, NaCl is reabsorbed to form hypotonic sweat. Modified from Rothman (1) and Sato (2).

2.2. Control of sweat secretion

The eccrine sweat glands respond to thermal stress through sympathetic cholinergic stimulation, and secretion is blocked by atropine. The sweat glands also possess α and β adrenergic receptors. In the cat, the concomitant release of vasoactive intestinal polypeptide (VIP) and acetylcholine from the same postganglionic fibers innervating eccrine glands has been reported. This finding might explain the cutaneous vasodilation seen during sweat secretion that enables the high blood supply needed for sweat secretion.

Thermoregulatory sweating increases with elevation in core temperature. At the initial stage, the number of recruited sweat glands is increased and, then, sweat secretion per gland increases to its maximal level. The sweat rate is modulated both by the mean skin temperature of the body and the local skin temperature. An increase in core temperature is about nine times more potent stimulus than an increase in mean skin temperature, while local skin temperature acts as a multiplier of the central control signal which determines the local sweat rate (see Chap. 6).

Another factor which may affect sweat rate is hidromeiosis. After about 1 hour of profuse sweating, a reduction in sweat rate is often observed, especially in moist heat. Previously, this reduction in sweat rate was considered to be a result of glandular fatigue. However, it has also been observed with wetting of the skin surface, while sweat rate is recovered with drying of the skin. It does not occur in dry heat or on areas of the skin stripped of the stratum corneum. These findings suggest that the cause of hidromeiosis is the obstruction of sweat ducts by swelling of the stratum corneum.

2.3. Sweat composition

The eccrine sweat glands secrete a dilute sweat containing only 0.5-1% of solids. One-half of the solids are inorganic salts, and the remaining half organic substances. Most of the inorganic salt is NaCl. However, the concentration of NaCl shows a large individual variation, with a range of 20 to 120 $mM \cdot l^{-1}$. This concentration increases with sweat rate because the proportion of these ions reabsorbed from the isotonic precursor sweat decreases before reaching the body surface. The maximum level of NaCl in sweat decreases with heat acclimatization, however, due to an increase of Na^+ reabsorption (see Chap. 11.1). The sweat NaCl concentration has important implications regarding water balance. Individuals with lower sweat NaCl concentration show a higher increase in blood osmolality than those with higher concentration because more of the ions are reabsorbed from the isotonic precursor sweat. Thus, the high rate of NaCl reabsorption elevates the osmolality of body fluids and causes fluid to shift from intracellular to extracellular compartments. It also stimulates drinking behavior after sweating.

Other major substances lost in the sweat are urea, lactic acid, and potassium ion. When the sweat rate is low, the concentrations of these substances are higher in sweat than those in the plasma. When the sweat rate is high, the concentrations of these constituents decrease. However, the role of the sweat gland as an excretory organ for these substances is minor; a rather high sweat rate is required to achieve a significant amount of excretion.

3. Non-Evaporative Heat Transfer and Control of Skin Blood Flow

Non-evaporative or "dry" heat transfer takes place through the channels of radiation, convection and conduction as the result of temperature differences between the body surface and the environment. The biophysics of these phenomena were described in Chap. 2. These heat transfers are modulated by autonomic regulation and also by behavior. Heat from the deep body tissues is transferred to the skin surface largely by the blood flow. Autonomic regulation of skin blood flow therefore controls the extent of heat loss to the environment. Behavioral reduction of the effective body surface for radiation, convection and conduction is observed in cold environments, while an increase in body surface area available for such heat exchange is observed in hot conditions.

3.1. Radiation

Radiant heat takes into account short wave radiations from the sun and infrared radiation transfer between the body and its surroundings. Human skin reflects about 40% and absorbs about 60% of the incident solar radiation of shorter wavelengths, including the visible and ultraviolet spectra. Long-wave, nonpenetrating infrared radiation is almost totally absorbed at the surface of the skin and is involved in the heat exchange with the surroundings. Long-wave radiation covers a range of wavelengths that has a maximum energy at a wavelength of $\lambda = 2{,}897/T$ µm, where T is the absolute temperature. Thus, skin at a temperature of 30 °C radiates infrared heat at a peak wavelength of 9.6 µm. However, heat rays are also radiated from the walls and other objects in the immediate environment toward the body. If the temperature of the body surface is greater than that of the surroundings, a greater quantity of heat is radiated from than to the body. A person at a thermoneutral temperature loses about 60% of the total heat loss by radiation. The magnitude of radiative heat transfer is a function of skin temperature, and the effective radiating area of the body achieved by behavior.

3.2. Conduction and convection

Body heat is lost by conduction to air and other objects in contact with skin. The kinetic energy of the molecular motion of the skin is determined by temperature; energy is transferred to the molecules in contact with the skin if the objects are colder than the skin. The thermal diffusivity is determined by the ratio of the thermal conductivity to the thermal capacity. Air has a very low thermal conductivity (0.024 W/m·K); this has important implications in the regulation of body temperature. When air warmed to skin temperature is covered with clothing, fur, or by piloerection, the amount of heat loss from the body is reduced considerably because of the low thermal conductivity of air. On the other hand, water has a high thermal conductivity (0.561 W/m·K), more than 20 times that of air, and a high heat capacity; thus, heat loss is much higher in water than in air. Helium is used instead of nitrogen in hyperbaric chambers to avoid nitrogen narcosis, and in this case, heat loss from the skin increases due to the high thermal conductivity of this gas (14.22 W/m·K). In humans, heat loss from the body to other objects which are in contact with the skin is rather small because the body surface is covered with clothing and the air layer beneath the clothing. In contrast, it is increased when the body is in contact with objects of high thermal conductivity such as soil or copper (403 W/m·K). This channel of heat loss is used by animals. An example is lying on a cement floor when hot to regulate body temperature through this behavioral mechanism. A small quantity of heat is also transferred to ingested food and water; ingestion of cool water suppresses the elevation of body temperature, especially during exercise in heat.

Heat loss by conduction to air ceases when the temperature of the air adjacent to the skin equals that of the skin. When the heated air is carried away from the body surface by convection, new unheated air is brought in contact with the skin, making possible additional heat loss.

Convective heat exchange depends on the redistribution of molecules within air or fluid in contact with the skin surface. Natural or free convection occurs when air in contact with skin is warmed and gives rise to buoyancy forces. Forced convection of air movement by a fan, wind or body movement increases convective heat exchange by disrupting the boundary air layer. Clothing modifies the convective heat exchange, limiting the movement of heated air adjacent to the skin. This has an important impact on the regulation of body temperature.

The heat loss from the body by conduction and convection is about 15 % of the total loss of body heat in humans in temperate conditions.

3.3. Control of skin blood flow and skin temperature

As outlined previously, each mode of heat transfer depends on the difference between the skin surface and environmental temperature. Subcutaneous fat works well as an insulator in the cold. But it offers little thermal insulation in hot conditions since heat transfer from the core to the skin is mainly dependent on the blood flow, which by-passes the fat depots.

(i) Control of vascular tone

During cold exposure, cutaneous blood flow, especially in uncovered skin areas such as the hands, feet, lips, ears and nose, is reduced almost entirely by the action of the sympathetic nervous system in response to changes in internal body or environmental temperature. The vasodilation in these regions during heat exposure largely occurs as the result of the withdrawal of this vasoconstrictor activity. The increase in blood flow is about 60 fold in the fingers and about 30 fold in the hands, while it is about 7 fold in the proximal portions of the extremities and trunk. Changes in blood flow of the fingers are especially effective for the regulation of heat loss because their surface area per unit mass is high. The arterioles and arteriovenous anastomoses that supply blood to the venous plexus of the skin are innervated by adrenergic fibers and possess α_1-receptors. Up to about 30 °C of environmental temperature, modulation of vasoconstrictor nerve activity suffices to regulate heat loss to maintain body temperature within a normal range. During heat stress, cutaneous circulation of the extremities and trunk is increased by active vasodilation.
The mechanism of active cutaneous vasodilator system remains unsolved; links with sudomotor control have been suggested, e.g., bradykinin, VIP, etc.

The blood vessels of the skin are constricted by noradrenaline. The action of noradrenaline varies with temperature. The contractile response of vascular smooth muscle at 30 °C is about 5 fold greater than it is at 41 °C. With further lowering of temperature, the response becomes attenuated such that, at 15 °C, the magnitude of the vasoconstriction is similar to that observed at 41 °C. At temperatures lower than 10 °C, the effect of noradrenaline is not observed, presumably due to a change in the affinity of adrenergic receptors. Local heating of finger tips causes vasodilation, while local cooling by water less than 10 to 18 °C causes vasoconstriction. In the cold, local vessels show a cyclic vasodilation, called the Lewis reaction or cold-induced vasodilation (CIVD). It has been suggested that CIVD is a result of the cyclic loss of the responsiveness of the vascular smooth muscle to noradrenaline that occurs with the lowering of local temperature, perhaps due to local tissue hypoxia and hypercapnia. With this loss of responsiveness, the blood flow is increased. This results in the warming of the local temperature until, again, vasoconstriction is observed. This cyclic change in finger tip circulation is advantageous in limiting heat loss and preventing frostbite.

(ii) Anatomical arrangement of blood vessels and heat exchange

As discussed in the previous section, skin blood vessels in uncovered skin are effective in controlling blood flow. This control is achieved by a specific vascular structure, the arterio-venous anastomoses (AVAs). The AVAs are short vessels with muscular walls that join an artery to a vein and are richly innervated with sympathetic nerve fibers. When they are open, blood short-circuits the route through the capillaries, increasing blood flow and local temperature.

B-10

In a cool environment, the arterial supply of the arm in man lies in close proximity to veins which drain this extremity, and warm arterial blood entering the limb is cooled by the returning, colder, venous blood. As a result, skin temperature is lowered and heat is conserved to maintain core temperature during exposure to a cold environment. In a hot environment, blood returns to the core via the surface veins and the skin temperature approaches that of the core temperature so that heat loss is maintained. This heat exchange between arteries and veins running in parallel is called countercurrent heat exchange and is well developed in certain species. In the appendages of whales, for example, the artery runs concentrically inside trabecular veins; this conserves heat that would otherwise be lost to the surrounding cold sea water.

A similar heat exchange function is served by the carotid rete. In some species, the blood supply to the hypothalamus runs via a rete of arterial vessels. This rete lies within a venous sinus that receives blood from the nasal passage, where it has been cooled. The arterial blood on its way to the brain passes through a heat exchanger so that the hypothalamus is protected from an excessively high temperature. Protection of the brain against high temperature by selective brain cooling has also been suggested to occur in humans, albeit without a rete.

4. The Effect of Heat Loss Mechanisms on Body Fluid and Circulation

Heat loss mechanisms rely on the change of skin blood flow and evaporation of body fluid, which means that alterations in circulation and body fluid take place as a consequence of thermoregulatory demands, which, in turn, means that these functions are limiting factors in the maintenance of body temperature. For example, sweating causes hyperosmolality of body fluids and hypovolemia, thus reducing both sweating and cutaneous vasodilation, hence impairing thermoregulation. The redistribution of blood flow for thermoregulation causes a reduction of central venous pressure.

B-11

During passive heating, these regulatory responses can maintain circulatory function up to about 40° C, while skin blood flow is restricted at body temperatures higher than this. Under conditions which simultaneously require increased blood flow to other beds, such as to the muscles during exercise in the heat, skin blood flow is maximal at about 38° C of body temperature, limiting heat flow to the skin and thus causing further elevation of body temperature. These findings suggest a hierarchic structure of the homeostatic mechanisms related to thermoregulation, body fluid homeostasis, and circulation. The central mechanism controlling this subtle balance between body fluid, body temperature and circulation has not been elucidated sufficiently as yet (see also Chap. 12).

5. Summary

Heat loss to maintain human body core temperature is regulated by behavioral as well as physiological mechanisms. The physiological responses to a heat load include cutaneous vasodilation to transfer heat from the core to the body surface, and to a cold load cutaneous vasoconstriction to conserve heat in the core and limit its loss from the body surface. These vasomotor responses are controlled by the autonomic nervous system according to the levels of core and skin temperatures. Behavioral temperature regulation modifies this physiological thermoregulation to minimize thermal discomfort. So long as skin temperature is lower than ambient temperature, the dissipation of body heat can occur by non-evaporative heat loss, i.e., by radiation, convection and conduction. In hot environments, the heat gradient between the skin and environment being reduced, the dependency on evaporative heat loss is proportionately increased. In humans, the sweating mechanism is well developed, and the tolerance to heat is high. During exercise, especially in extreme heat, both of these themoregulatory responses have important impacts on cardiovascular function and body fluid homeostasis such that heat-induced changes in circulatory function and dehydration may competitively limit thermoregulatory responses.

6. References
1. Rothman S (1954) *Physiology and Biochemistry of the skin.* Univ. Chicago Press, P. 154.
2. Sato K (1993) The mechanism of eccrine sweat secretion. In: Gisolfi CV, Lamb DR, Nadel ER (eds), *Perspectives in Exercise Science and Sports Medicine. Vol. 6: Exercise, Heat, and Thermoregulation.* Benchmark Press, Indianapolis, pp. 85-110.

Suggested readings for Chapter 5

Cabanac M (1995) *Human Selective Brain Cooling*. Landes, Austin.

Gagge AP, Gonzalez RR (1996) Mechanisms of heat exchange: biophysics and physiology. In: Fregly, MJ and Blatteis CM (eds), *Handbook of Physiology Section 4: Environmental Physiology, Vol. 1*, Oxford University Press, New York, pp. 45-84.

Johnson JM, Proppe, DW (1996) Cardiovascular adjustments to heat stress. In: Fregly, MJ and Blatteis CM (eds), *Handbook of Physiology Section 4: Environmental Physiology, Vol. 1*, Oxford University Press, New York, pp. 215-243.

Sato K (1993) The mechanism of eccrine sweat secretion. In: Gisolfi CV, Lamb DR, Nadel ER (eds), *Perspectives in Exercise Science and Sports Medicine. Vol. 6: Exercise, Heat, and Thermoregulation*. Benchmark Press, Indianapolis, pp. 85-110.

Sawka MN, Wenger CB (1988) Physiological responses to acute exercise-heat stress. In: Pandolf, KB, Sawka, MN, Gonzalez, RR (eds), *Human Performance Physiology and Environmental Medicine at Terrestrial Extremes*. Benchmark Press, Indianapolis, pp. 97-151.

Self-evaluation questions for Chapter 5

True or false
1. When heat loss from the body is blocked completely, the body temperature of a 70 kg man with a heat production of 10,000 kJ·day^{-1} will increase from 37 to 70 °C. (Use 3.47 kJ·kg^{-1}·°C^{-1} for thermal capacity of the human body and 20% efficiency for metabolism.)
2. Humans do not use behavioral temperature regulation.
3. More than 95% of heat is transferred from the core to the surface of the body by circulation.
4. Countercurrent heat exchange is an effective mechanism for increasing blood flow to the limbs.
5. Heat transfer between the sun and earth is due to radiation by long-wave infrared heat rays.
6. Radiative heat exchange does not occur when the environmental temperature is greater than the body temperature.
7. Evaporative heat loss is absent under comfortable environmental conditions.
8. When a 70 kg man is exercising at a rate of 600 kJ·h^{-1}, about 200 g of sweat is enough to maintain a steady-state body temperature.
9. A wet suit is very effective in preventing heat loss by convection.
10. The survival time of one kept afloat by a life jacket in water at 15 °C is likely to be 2-3 hours.

Chapter 6

Learning Objectives
1. Basic cellular mechanisms of temperature sensitivity in cutaneous receptors and hypothalamic neurons.
2. Neural pathways for afferent thermal information ascending from the skin to the hypothalamus.
3. Hypothalamic integration of central and peripheral thermal information.
4. Endogenous factors (pyrogens, plasma osmolality, glucose concentration, reproductive hormones, circadian time) that affect hypothalamic thermoregulatory neurons.

Bullets
- B-1 Cutaneous receptors relay thermal information to the hypothalamus.
- B-2 Hypothalamic neurons receive ascending somatosensory signals.
- B-3 Cold receptors code thermal and relay thermal information as dynamic firing rate, static firing rate, and bursting activity.
- B-4 Ionic conductances account for the thermosensitivity of cutaneous receptors.
- B-5 Depolarizing prepotentials contribute to the thermosensitivity of hypothalamic warm-sensitive neurons.
- B-6 Cold-sensitive neurons are synaptically inhibited by warm sensitive neurons.
- B-7 Pyrogens affect thermosensitive neurons.
- B-8 Preoptic warm-sensitive neurons receive afferent synaptic input from skin thermoreceptive pathways.
- B-9 Hormones and other endogenous factors affect hypothalamic thermosensitive neurons.
- B-10 Thermosensitive hypothalamic neurons are affected by an internal circadian clock.

Chapter 6

NEURAL THERMAL RECEPTION AND REGULATION OF BODY TEMPERATURE

JACK A. BOULANT
Department of Physiology, Ohio State University
Columbus, Ohio, USA

1. Introduction

Regulation of body temperature depends on the nervous system's ability to sense and integrate thermal information from the external environment and deep within the body core. Somatosensory pathways provide communication between peripheral thermoreceptors and central neurons; and the integrative capacity of hypothalamic neurons to process this information is a hallmark of thermoregulation.

Environmental temperature is sensed by thermoreceptors in the skin. These cutaneous thermoreceptors are free nerve endings which produce action potentials in their small diameter, unmyelinated or poorly myelinated afferent fibers (i.e., C fibers). There are two basic types of cutaneous thermoreceptors, viz., warm receptors and cold receptors. For humans in a thermal neutral environment, such as a comfortable room temperature, the mean skin temperature is about 33°C. Many cutaneous thermoreceptors display spontaneous firing rates at 33°C. Firing rate is the receptor's discharge frequency, expressed as impulses (or action potentials) per second (i.e., imp/sec). Cutaneous warm receptors increase their firing rates during moderate warming and decrease their firing rates during cooling; and conversely, cold receptors increase their firing rates during moderate cooling and decrease their firing rates during warming. Since many receptors are active near 33°C, both types of receptors can relay information about warming and cooling by either increasing or decreasing their firing rates.

The cutaneous receptors' afferent fibers synapse on dorsal horn spinal neurons, which send their axons centrally in the lateral spinothalamic tract. Some of these axons project to the ventrobasal thalamus, where thalamic neurons relay this thermal information to the somatosensory cortex. In the thalamus and cortex, thermal and tactile information is processed, allowing us to perceive exact temperatures at precise locations on the skin surface. This information, however, is not useful in the hypothalamic regulation of body temperature, which requires information about average temperature of large skin areas. For this information, the same dorsal horn spinal neurons send axons (possibly collateral axons) to neurons in the brain stem reticular formation. Converging input from different spinal neurons allows brain stem neurons to sense temperature over relatively large portions of the skin surface. The exact site of convergence in the reticular formation is not well known and remains a matter of speculation. Nevertheless, it is likely that reticular neurons relay the peripheral thermal information forward to neurons in and near the rostral hypothalamus, especially the preoptic and anterior hypothalamus (PO/AH).

Approximately 30-40% of PO/AH neurons are, themselves, thermosensitive and respond to both increases and decreases in their own temperature. During exercise, for example, the elevated core body temperature warms the hypothalamus by the warmed blood that perfuses this region. The PO/AH thermosensitive neurons sense this temperature increase and change their own firing rates. In addition to sensing their own, local temperature, most of the PO/AH thermosensitive neurons receive synaptic inputs from afferent pathways (e.g., from reticular formation neurons) that relay information about temperature in the skin and other locations throughout the body (e.g., spinal cord, deep blood vessels, gastrointestinal tract). The PO/AH temperature sensitive neurons send some of their dendrites laterally to receive synaptic input from ascending and descending fibers in the median forebrain bundle that runs throughout the lateral hypothalamus. In addition, these same PO/AH neurons send some of their dendrites medially to receive other afferent signals from periventricular fibers or, possibly, to receive chemosensitive signals from endogenous substances in the cerebrospinal fluid of the third ventricle. These PO/AH neurons project their axons to different locations throughout the nervous system to control a variety of autonomic and behavioral thermoregulatory responses. Thus, the thermosensitive PO/AH neurons have the ability to compare and integrate central and peripheral thermal information. As a result of this integration, PO/AH neurons produce efferent signals to elicit the thermoregulatory responses that are most appropriate for the given internal and external environmental conditions. Consequently, body core temperature is usually maintained relatively constant under a variety of conditions.

2. Thermoreceptors and Neuronal Thermosensitivity

2.1. Cutaneous thermoreceptors

As indicated above, cutaneous warm receptors increase their firing rate during increases in skin temperature, and cutaneous cold receptors increase their firing rates during decreases in skin temperature. Like most receptors, cutaneous thermal receptors display adaptation and have both dynamic and static firing rates. Figure 1 shows the action potential activity of a cold receptor that is held, initially, at a constant temperature of 34°C before being transiently cooled to different temperatures (ranging from 31°C to 24°C) and finally returned back to 34°C. At the beginning of each record, at 34°C, the cold receptor has a low spontaneous firing rate. When the receptor is suddenly cooled a few degrees, it shows a large increase in firing rate (i.e., dynamic firing rate) which rapidly adapts over time to a new steady-state firing rate (i.e., static firing rate). Greater cooling increases both the dynamic and static firing rates.

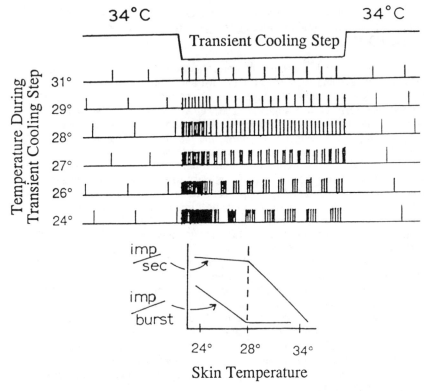

Figure 1. Effect of transient cooling on the firing rate of a cutaneous cold receptor. Each trace shows the receptor's action potential activity when temperature is briefly changed from a constant 34°C to a cooler temperature (ranging from 31°C to 24°C). Following each transient cooling, the temperature is returned back to 34°C. The receptor has a low spontaneous firing rate at 34°C. During each of the cooling steps, the receptor shows a large dynamic firing rate and rapidly adapts to a new static or steady-state firing rate. Greater cooling increases both the dynamic and static firing rates. When the temperature is cooled below 28°C, the static firing rates display bursting activity.

In addition, during even greater cooling, the static firing rates of some cold receptors show "bursting" activity. This allows cold receptors to code thermal information even after the static firing rate has reached maximal long-term levels. Instead of coding information as action potentials per second (i.e., imp/sec), bursting activity codes information as action potentials per burst (i.e., imp/burst). In Figure 1, for instance, the cold receptor might reach a maximum steady-state firing rate of 10 imp/sec if it is cooled 6° (i.e., from 34°C to 28°C). If this same receptor is cooled further (e.g., 7°), it may still fire at approximately 10 imp/sec; however, instead of single, periodic action potentials, it may fire a burst of 2 action potentials and then have an interval separating another burst of 2 action potentials. Each second, the receptor may have five bursts with 2 action potentials in each burst. Therefore, it still maintains an overall firing rate near 10 imp/sec. If the same receptor is cooled even more, to 24°C, it can increase its bursting activity while maintaining a fairly constant firing rate. For example, it may have 5 action potentials in each burst, but in this case, during each second, it might fire only two bursts. This bursting activity allows cold receptors to continue to code thermal information beyond the range in which a maximum long-term firing rate is reached. Thus, cutaneous cold receptors code thermal information by dynamic firing rate and static firing rate, and (as shown in the graphs at the bottom of Figure 1) the static firing rate codes the information by action potentials/second and by action potentials/burst.

The exact mechanisms for the thermosensitivity of cutaneous cold and warm receptors are not known; however, many believe that they are similar to those of large invertebrate neurons, such as sea hares and snails. A primary mechanism for cold-sensitivity is believed to be the metabolically-driven sodium-potassium (Na^+-K^+) pump. Near room temperature, the Na-K pump transports sodium ions out of a cell and potassium ions into a cell. This electrogenic pump produces a gradient that allows positive K^+ ions to leak out of the cell. This hyperpolarizes the cell's membrane and accounts for much of the cell's negative resting membrane potential. Warming accelerates the Na-K pump. This hyperpolarizes the membrane, and causes cold receptors to decrease their firing rates. On the other hand, cooling slows the Na-K pump which allows cold receptors to depolarize, producing an increased firing rate.

The thermosensitivity of cutaneous warm receptors, on the other hand, may be due to warm-induced depolarization associated with the effect of temperature on the ratio of Na^+ and K^+ permeabilities. In certain cells, temperature may have a relatively greater effect on passive Na^+ currents, compared to K^+ currents. Thus, with warming, there would be relatively more depolarizing inward Na^+ current compared to the hyperpolarizing outward K^+ current. This depolarization would result in an increased firing rate with warming.

2.2. Hypothalamic thermosensitive neurons

In the PO/AH and in most other hypothalamic areas, about 30% of the spontaneously firing neurons are warm-sensitive, while less than 10% of the neurons are cold-sensitive. During increases in hypothalamic temperature, warm-sensitive neurons increase their firing rates (see Figure 2), and cold-sensitive neurons decrease their firing rates. The remaining majority of neurons show little or no change in their firing rates during warming or cooling; these neurons are classified as temperature-insensitive. Although cutaneous thermoreceptors are associated with thermal-induced depolarization, the membrane potentials of hypothalamic neurons are not dramatically affected by temperature changes. Thus, the mechanisms of thermosensitivity appear to be different in central and peripheral thermoreceptors.

As shown in Fig. 2, at neutral body temperatures of 36°-37°C, PO/AH warm-sensitive neurons often have spontaneous firing rates with constant intervals between the action potentials (i.e., interspike intervals). The neuronal firing rates increase with warming and decrease with cooling. In addition, like all other neurons, action potential amplitudes vary inversely with temperature. Figure 2A shows the action potentials produced during one-second records taken at 32°, 36°, and 39°C. Figure 2B shows the membrane potentials during the interspike intervals at these 3 temperatures. Following an action potential, warm-sensitive neurons display a pacemaker potential or depolarizing prepotential, such that the membrane

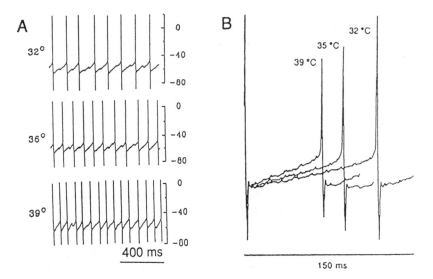

Figure 2. Effect of temperature on the activity of a warm-sensitive neuron recorded intracellularly, *in vitro*, in a rat preoptic tissue slice.
A. Action potentials occurring during one-second intervals at three different temperatures. The spontaneous action potentials are preceded by depolarizing prepotentials. When the depolarization reaches threshold, an action potential is produced. Warming increases the rate of depolarization in the prepotentials, and this increases firing rate.
B. Effect of the three temperatures on the interspike interval for the same warm-sensitive neuron. The superimposed records show individual action potentials and the subsequent depolarizing prepotential and action potential. Warming increases the rate of rise of the prepotential, causing a decrease in the interspike interval which increases the firing rate. (Modified from Griffin, Kaple, Chow & Boulant, *J. Physiol. (London)*, 492:231-242, 1996.)

slowly depolarizes until threshold is reached and the next action potential is produced. Temperature affects the rate at which this prepotential depolarizes, and this appears to be a primary mechanism for neuronal warm-sensitivity. Figure 2B shows that warming increases the rate of depolarization, allowing the next action potential to occur sooner. This warm-induced shortening of the interspike interval causes the neuron's firing rate to increase.

Much of the depolarizing prepotential appears to be due to a decrease in transient outward hyperpolarizing currents, such as the potassium A current (I_A). I_A is a brief outward movement of K^+ ions that helps maintain a negative membrane potential during the time period between action potentials. After an action potential, I_A activation tends to hyperpolarize the membrane potential. Following this, I_A inactivation occurs, which allows the membrane to slowly depolarize toward threshold. I_A inactivation is temperature-dependent. Cooling slows the rate of I_A inactivation and, thus, slows the depolarizing prepotential and firing rate. Conversely, warming increases the rate of I_A inactivation, and this increases the prepotential's rate of depolarization, leading to an increased firing rate. This effect of warming is shown in the 39° example in Figure 2B.

While PO/AH warm-sensitive neurons appear to be intrinsically thermosensitive, this is not the case for most PO/AH cold-sensitive neurons. When synaptic activity is blocked, warm-sensitive neurons remain thermosensitive, but thermosensitivity is lost in most cold-sensitive neurons. Furthermore, recent intracellular recordings suggest that PO/AH cold-sensitive neurons are highly dependent on inhibitory and excitatory synaptic inputs from nearby neurons. For these reasons, many believe that in the PO/AH, cold-sensitive neurons are spontaneously firing neurons that are synaptically inhibited by warm-sensitive neurons. These cold-sensitive neurons, therefore, increase their firing rates during cooling, because the inhibitory warm-sensitive neurons decrease their synaptic inhibition. Conversely, during warming, the increased firing rates in the warm-sensitive neurons cause the synaptically-inhibited neurons to decrease their firing rates, and thus, the inhibited neurons appear to be cold-sensitive.

3. Integration: Central and Peripheral Temperature

Figure 3 shows a neuronal model that describes synaptic connections in PO/AH neurons and how these neurons might control thermoregulatory responses. The model shows a cold-sensitive neuron (C) that receives synaptic inhibition from a warm-sensitive neuron (W). The cold-sensitive neuron may, itself, be spontaneously active; however, in support of recent intracellular studies, the model indicates that the cold-sensitive neuron's firing rate may also be driven by excitatory synaptic input from a temperature-insensitive neuron (I). During hypothalamic cooling, therefore, the amount of excitation (from neuron I) remains constant, but the amount of inhibition (from neuron W) decreases, and this causes the firing rate (in neuron C) to increase.

Unlike temperature-insensitive neurons, many PO/AH thermosensitive neurons are affected by pyrogens that cause fever by suppressing heat loss responses and evoking heat production and heat retention responses. Pyrogens often inhibit warm-sensitive neurons, excite cold-sensitive neurons, and have little effect on temperature-insensitive neurons.

B-7

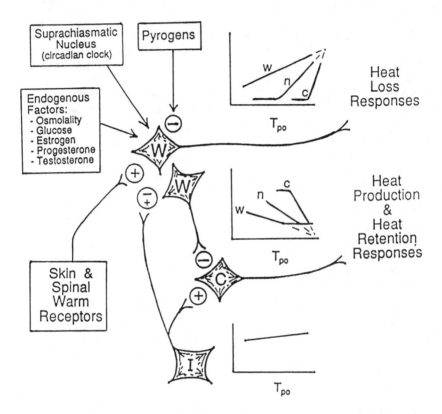

Figure 3. Neuronal model describing synaptic connections in three different types of neurons in the preoptic/anterior hypothalamus; i.e., warm-sensitive neurons (W), cold-sensitive neurons (C), and temperature-insensitive neurons (I); (+) indicates excitation, and (-) inhibition. Graphs on the right show changes in neuronal firing rates and changes in thermoregulatory responses during changes in preoptic temperature, T_{po}. Responses to T_{po} are shown at three skin temperatures: warm (w), neutral (n), and cold (c) skin temperatures.

Unlike temperature-insensitive neurons, PO/AH thermosensitive neurons receive much ascending input from thermoreceptors in the skin and spinal cord. In rabbits, the firing rates of approximately 60-70% of PO/AH warm-sensitive and cold-sensitive neurons are affected by changes in either skin or spinal cord temperatures,

B-8

and most of these neurons respond the same way to both central and peripheral temperature. For example, the PO/AH warm-sensitive neurons in Figure 3 might be expected to increase their firing rates in response to warming either the hypothalamus or the skin or the spinal cord; and, of course, the greatest increase in firing rate would occur during whole-body warming when all three temperatures increase.

The model in Figure 3 suggests that PO/AH thermosensitive neurons not only sense changes in core temperature, but also integrate central and peripheral thermal information to elicit the most appropriate responses for the given internal and external thermal conditions. When either hypothalamic or peripheral temperature is changed, the responses of many PO/AH thermosensitive neurons are similar to physiological and behavioral thermoregulatory responses. Because of these similarities, models like Figure 3 often present a simplified view suggesting that many PO/AH warm-sensitive neurons control heat loss (e.g., evaporative heat loss due to panting or sweating), while many PO/AH cold-sensitive neurons control heat production (e.g., shivering and nonshivering thermogenesis) and heat retention (e.g., cutaneous vasoconstriction and thermoregulatory behaviors). It should be reiterated that this model is an oversimplification, and it is possible that both warm- and cold-sensitive neurons control all types of thermoregulatory responses.

The top right graph in Figure 3 shows the changes in the firing rates of warm-sensitive neurons and the changes in heat loss responses during changes in PO/AH temperature, T_{po}. These responses to T_{po} are shown at three different skin temperatures: warm (w), neutral (n), and cold (c) skin temperatures. As the graph indicates, skin temperature changes the T_{po} set-point or the threshold temperature that must be reached before there is a change in the firing rate or heat loss response. In addition, skin temperature often alters the slope or sensitivity of these responses to changes in PO/AH temperature. Skin warming decreases the PO/AH set-point temperature and causes heat loss to occur even at a normal T_{po} (e.g., 37°c). Skin warming can also decrease the slope or the sensitivity of this response to a change in T_{po}. On the other hand, skin cooling increases the PO/AH set-point temperature and suppresses heat loss at temperatures above a normal T_{po}. Once set-point is exceeded, however, skin cooling can increase the slope or sensitivity of this response to a change in T_{po}.

The integrative responses of cold-sensitive neurons and heat production-heat retention are shown in the middle right graph in Figure 3. Heat production increases when T_{po} decreases below a PO/AH set-point temperature, and the firing rates of many cold-sensitive neurons show similar responses. Skin cooling increases both T_{po} set-point and PO/AH thermosensitivity; while skin warm decreases both T_{po} set-point and PO/AH thermosensitivity.

The responses in cold-sensitive neurons and heat production are the mirror-images of the responses seen in warm-sensitive neurons and heat loss. This fact supports the model's suggestion that cold-sensitivity is due to synaptic inhibition from warm-sensitive neurons. During skin warming, the increased firing rate in the warm-sensitive neuron causes a decreased firing rate in the cold-sensitive neuron; but the thermosensitivity of the cold-sensitive neuron is also reduced because the thermosensitivity of the warm-sensitive neuron is reduced. During skin cooling, the decreased firing rate in the warm-sensitive neuron causes an increased firing rate in the cold-sensitive neuron; but the thermosensitivity of the cold-sensitive neuron is increased because the thermosensitivity of the warm-sensitive neuron is increased.

4. Integration: Thermal and Non-Thermal Information

In addition to pyrogens, Figure 3 indicates that PO/AH thermosensitive neurons can be influenced, both directly and synaptically, by a variety of other endogenous factors, including reproductive hormones. While not shown in the model, these factors affect both temperature-sensitive and -insensitive neurons. The sensitivity of PO/AH neurons to non-thermal factors can account for many of the strong interactions between thermoregulation and other homeostatic systems. As an example, the changes in body temperature that occur in women during the menstrual cycle are likely due to changes in progesterone and estrogen levels. Just before ovulation, there is often a brief drop in body temperature that is associated with rising estrogen levels. During ovulation and in the days that follow, there is an elevated body temperature associated with high progesterone levels.

Figure 3 indicates that PO/AH thermosensitive neurons are also affected by plasma osmolality and glucose concentration. These endogenous factors also affect body temperature. For example, thermoregulation is impaired during increases in plasma osmolality (e.g., during dehydration) or during decreases in plasma glucose; and it is likely that some of these thermoregulatory changes are due to direct effects on the activity of hypothalamic neurons.

While endogenous factors affect thermoregulation, body temperature also affects many of the non-thermal regulatory systems. For example, increases in temperature tend to suppress feeding but increase water retention mechanisms (i.e., causing increased drinking and antidiuretic hormone secretion). It is likely, therefore, that both PO/AH temperature-sensitive and -insensitive neurons play a role not only in thermoregulation, but also in the regulation of a variety of homeostatic systems. This would explain the close relationship between the regulation of body temperature and other homeostatic systems.

Finally, Figure 3 indicates that PO/AH neurons receive information from the body's circadian clock, which lies in an adjacent area of the hypothalamus, the suprachiasmatic nucleus (SCN). In the rat, a nocturnal animal, the firing rates of SCN neurons show an intrinsic circadian activity, with higher firing rates in the day and lower firing rates at night. This circadian activity occurs even *in vitro*, in hypothalamic tissue slices. SCN circadian information can be relayed synaptically to nearby PO/AH neurons; and it is likely that this accounts for circadian variations in thermoregulation, where (in humans) body temperature is regulated higher during the day-time compared to the night-time. Moreover, if PO/AH temperature-sensitive and -insensitive neurons control both thermal and non-thermal regulatory systems, then the SCN input to the PO/AH may account for the circadian rhythms observed in a variety of homeostatic systems.

5. Summary

The mechanisms of thermosensitivity differ between cutaneous warm and cold receptors and hypothalamic thermosensitive neurons. In peripheral receptors, warming or cooling can depolarize free nerve endings to produce increases in firing rate. In hypothalamic neurons, however, warm-sensitivity is dependent on thermally-induced changes in the rate of rise in depolarizing prepotentials that determine the interval between successive action potentials. Many hypothalamic cold-sensitive neurons are not intrinsically thermosensitive, but rather depend on synaptic inhibition from nearby warm-sensitive neurons.

Cutaneous temperature receptors relay peripheral thermal information over somatosensory pathways to hypothalamic thermosensitive neurons that integrate this thermal information to control a variety of heat loss, heat retention, and heat production responses. These same hypothalamic neurons are also affected, both directly and synaptically, by endogenous factors (e.g., pyrogens, reproductive endocrines, osmolality, glucose, and the suprachiasmatic circadian clock). These non-thermal endogenous factors account for cyclic changes in body temperature and for changes in thermoregulation during fever and alterations in homeostatic conditions.

Acknowledgements

Professor Boulant's research is supported by NIH grant NS14644 and the Hitchcock Professorship in Environmental Physiology.

Suggested Readings

Boulant, J.A. 1996. Hypothalamic neurons regulating body temperature. In: APS Handbook of Physiology, Section 4: Environmental Physiology, ed. M.J. Fregly & C.M. Blatteis, pp. 105-126, Oxford University Press, New York.

Boulant, J.A., and N.L. Silva. 1989. Multisensory hypothalamic neurons may explain interactions among regulatory systems. News in Physiological Sciences 4: 245-248.

Boulant, J.A., M.C. Curras, and J.B. Dean. 1989. Neurophysiological aspects of thermo-regulation. In: Advances in Comparative and Environmental Physiology, 4. Animal Adaptation to Cold, edited by L.C.H. Wang. Berlin: Springer-Verlag, pp. 117-160.

Self-evaluation questions for chapter 6

1. Trace the ascent of thermal information from cutaneous thermoreceptors to the thermosensitive neurons in the preoptic-anterior hypothalamus.

2. Explain the cellular mechanisms of temperature sensitivity for both types of cutaneous thermoreceptors and both types of hypothalamic thermosensitive neurons.

3. Describe the different types of hypothalamic neurons and their possible roles in thermoregulatory responses.

4. Describe the firing rate responses of a cutaneous cold receptor to different degrees of cooling.

5. List the endogenous factors that can affect hypothalamic thermosensitive neurons, and describe how these factors might alter heat loss and heat production responses.

6. Describe the synaptic connections between hypothalamic warm-sensitive neurons, cold-sensitive neurons, and temperature-insensitive neurons.

Chapter 7

Learning Objectives
1. Understanding what is behavior.
2. Knowledge of the place of behavior in temperature regulation.
3. Basic knowledge of thermal comfort and its physiological signals.
4. Understanding situations, in health and disease, where motivations and ensuing behaviors clash with temperature regulation.
5. Understanding the reciprocal situations where a subject's temperature regulation is modified by pathology and the subject uses behavior to thermoregulate.

Bullets
B-1 Behavior takes many forms.
B-2 Behavior is physiology.
B-3 The seeking of pleasant temperatures achieves temperature regulation.
B-4 When other motivations rank above thermal comfort, the resulting behavior may be dangerous.
B-5 Alcohol is dangerous in cold environment.
B-6 Emotion elevates the set-point.
B-7 A new hypothesis regarding temperature regulation during anesthesia.
B-8 Behavior saves the lifes of erythrodermic patients.
B-9 Ageing and sexual differences should be examined in the light of selective brain cooling.
B-10 The set-point is adjustable.

Chapter 7

THERMIATRICS AND BEHAVIOR

MICHEL CABANAC
*Département de Physiologie, Faculté de Médecine
Université Laval, Québec, G1K 7P4 Canada*

1. Introduction

The importance of thermoregulatory behavior is often underestimated by the physician because both temperature regulation and behavior are often neglected in medical curricula. Therefore, when pathology strikes the physician is mostly dependent on his intuition. Yet, behavior is a thermoregulatory response. In the long term, behavior is even the only thermoregulatory response that allows humans to cope with hostile climates.

2. Thermoregulatory Behaviors

Thermoregulatory behavior consists of adjustments of the subject's situation to his environment. Behaviors may be ranked by order of complexity.

2.1 Postural Adaptation and Migration. Among the simplest behaviors, postural adaptation is already quite efficacious when applied to the flux of energy of solar radiation. In a dry climate, the sun irradiates about 1 kW·m^{-2}; about 1/10 of this power is received by a clothed human. A person can modulate by a factor of 3 the heat gained from solar heat flux just by body positioning, crouching or spreading limbs. Migration can be understood not only as continental crossing but also as more modest travel, as when a subject seeks protection from cold gale or solar beams.

2.2 The Building of Microenvironments and Operant Behavior [1]. This spares the locomotion needed by migration and, at the same time, satisfies the physiological needs. The human species tends to live in a totally artificial environment. Its natural environment is artefactual. As a result, the human species has been able to invade all latitudes by building thermal micro-climates. Heated houses provide favorable temperatures, but clothes and beds also slow down the exchange of heat between body and environment. As a result, humans live permanently in tropical conditions, even under extreme latitudes, and have started to invade outer space. A variation of building micro-climates is the use of operant behavior, which modifies the subject's immediate environment. Operant behavior consists of the use of apparatus such as heater, fan, or air-conditioner (Fig.1).

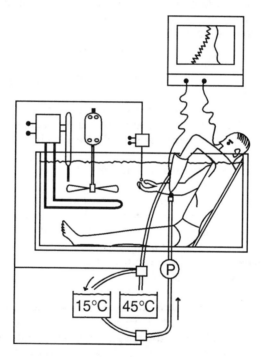

Fig.1 Example of an operant behavior to measure preferred skin temperature that also results in a minute change in the subject's micro-environment (his left hand only). The thermostated, stirred bath clamps the temperature of most of the subject's skin sensors. From within the bath, the subject can actuate the electric valves that direct cold or hot water from the tanks to the glove surrounding his left hand. P=pump. Equations 1 and 3-5 were obtained with methods of this type (from Cabanac, Massonnet & Belaiche, J. appl. Physiol. 1972, 33:699).

2.3 Social Behavior. When parents provide their offspring with thermal protection, they behaviorally ensure the environmental conditions for survival. In humans, the small body mass and large relative surface area of neonates render them prone to hypo- and hyperthermia. Thus, they rely mostly on behavior to thermoregulate. Their thermoregulatory behavior consists of an audible request for parental protection. Technology anticipates potential thermal imbalance and, through modelization, permits improved protection of the new-born. It should be noted that the therapy of patients suffering from hypothermia or hyperthermia also belongs in this category when a behavior beneficial to a patient is provided by another person.

2.4 Behavioral Self-Adjustment of the Subject. Behavior may be focused not only on the modulation of the environment, but also on the subject itself. In this case, behavior is quite similar to the autonomic response, and their separation may not be easy since the response lies within the body. The best example of such a response is thermoregulatory heat production from muscular exercise. Self-adjustment is not limited to the control of muscles. It has been demonstrated that subjects are able to modulate also their visceral function. The principle is that of operant conditioning. The subjects lower or enhance a given function, say heart rate, for the sake of obtaining the satisfaction of improving one's health. See [2] for review of examples of pathologies in the various behaviors just listed

3. Efficacy and Precision of Behavior

The few examples provided above show that behavior is both proteiform and ubiquitous in its service of temperature regulation. Behavior is the obligatory step in the human life process. This section will show that this servo-control is not only qualitative all-or-nothing, but is also precisely quantitative. Behavior is adjusted to the temperature needs of the body in the short-term as well as in the long-term.

B-2

3.1 Short-Term Adjustments. Strikingly similar equations describe the proximate thermoregulatory corrective responses, as functions of the body temperatures, whether the response is autonomic or behavioral (Table 1), where, R = the corrective response, T_{sk} = skin temperature, T_{set} = set-point temperature, T_b = core body temperature, and a and b are constants.

TABLE 1

BEHAVIORAL RESPONSE	$R = aT_{sk}(T_b - T_{set}) + b$	[3]	(1)
AUTONOMIC RESPONSE	$R = a(T_b - T_{set})$	[4]	(2)

3.2 Long-Term Adjustments [5]. The application of behavior to long-term physiological regulations has been developed recently by Mrosovsky.

In the long-term, humans would not be able to sustain their autonomic responses in hostile environments. Neither the amount of available water stored in the body, nor that of energy would allow prolonged sweating or shivering. Thus, long-term stay and survival in hostile climates, and even in a cool climate, would not be possible without behavior.

When behavior is not available for temperature regulation because it is used by a more urgent motivation, the autonomic responses take over the burden of maintaining a stable core temperature. Such a complementarity of behavioral and autonomic functions is reciprocal or mutual and can be found in pathological situations. When the autonomic response to environmental stress is hindered by disease or by therapeutic manoeuvres, the behavioral response takes over the regulatory process. This may give rise to another cost, behavioral this time, which is to be paid by the subject. Reciprocally, psychological or behavioral aggressions have somatic impacts. Stress is now viewed mainly as a general biological response to environmental demands, and experimental studies of stress always consist of behavioral hindrances which may interfere with temperature regulation. In the following, the reciprocal influences of behavior and pathology will be reviewed systematically. The link between a physiological need and the behavioral satisfaction of this need is a conscious signal: motivation. In the case of temperature regulation, this motivation is thermal comfort.

4. The Perception of Thermal Comfort

Thermal comfort, like other conscious perceptions, is a mental phenomenon; it is, therefore, a property of the brain. Three main types of information interact to produce thermal comfort: skin and core temperatures, and thermoregulatory set-point. These three inputs combine to produce the motivation for thermoregulatory behavior [6].

4.1 Skin Sensation. Skin temperature arouses a sensation. This obvious point requires no demonstration. Sensation is a quadridimensional event of consciousness. The first dimension is quality, which identifies the nature of the stimulus. The second dimension is magnitude, which identifies the intensity of the stimulus. The third dimension is affectivity (pleasure, displeasure); we shall see, below, that this dimension depends on inputs from the core temperature and the thermoregulatory set-point, i.e., this is where comfort and discomfort take place. The fourth dimension is time; it identifies the duration of the stimulus.

4.2 Influence of Core Temperature. The strong influence of core temperature on the perception of thermal comfort is evidenced by the reports of subjects whose core temperature is modified while their mean skin temperature remains stable. Thus, when core temperature deviates from normal, subjects complain of thermal discomfort although their skin temperature remains neutral. The primary component of such a discomfort is the displeasure aroused in the skin. To a hypothermic subject, cold skin feels unpleasant, but the same stimulus feels very pleasant to a hyperthermic subject. Thus, the pleasure or displeasure aroused by a

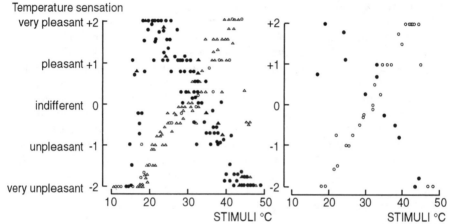

Fig.2 Left: affective responses to temperature stimuli between pain thresholds (15 and 45°C) given by a subject whose hand is stimulated for 30 s, and body is immersed in a bath similar to that of fig.1. Each dot is the response to a stimulus. Triangles, cold bath; circles, warm bath; the subject was hyperthermic, open dots, or hyperthermic, solid dots. It can be seen that core temperature determined the affective experience; e.g., to the hyperthermic subject, cold stimuli felt pleasant, and warm stimuli unpleasant.
Right: same legend as left except that in that case the subject's core temperature was the same (ca. 38°C) both with solid and open dots. Solid dots indicate the responses when the subject was feverish, and open dots the responses from control session without fever. It can be seen that the modified set-point was sufficient to change the subject's thermal preference, all other conditions being identical (from Cabanac, Physiol. Behav. 1969, 4:359).

thermal stimulus depends on the subject's internal state. This phenomenon is called *alliesthesia* (Fig.2). The internal signal that gives rise to an alliesthesic change is the difference between set temperature and actual temperature. Two models have been proposed to predict the preferred skin temperature (T_p):

$T_p = a(T_c-b)T_{mean\ sk}+c$ [3] \hfill (3),

$T_p = a+bT_c+cT_{mean\ sk}$ [7] \hfill (4),

and one predicting the subjective assessment, (SA):

$$SA = aT_c + b dT_{c/dt} + cT_{mean\ sk} + e dT_{mean\ sk/dt} + fS + g\ [8] \tag{5},$$

where a, b, c, e, f, and g are constants, T_c is core temperature, $T_{mean\ sk}$ is mean skin temperature, and S is a shivering factor (0 or 1). One model was proposed to predict alliesthesia:

$$a = f(T_c - T_{set})T_{mean\ sk,}\ [9] \tag{6}$$

where T_{set} is the set temperature of the biological thermostat. Alliesthesia occurs simultaneously in the various parts of the skin, which explains comfort and discomfort in various circumstances. For example, comfort persists when one side of the body is cooled and the other side warmed, or when the body receives asymmetrical radiation producing a difference of up to 13°C between front and back skin temperatures.

4.3 Set-Point. The signals recognized above as well as the proposed equations are those of temperature regulation. This regulation possesses an additional implicit signal: the set-point. Actual internal temperature is compared to the set-point 'wanted' by the organism. The activating signal for the regulatory responses, the 'error signal', is the difference between the actual temperature and the set-point. When an error signal is detected, the organism produces the available corrective responses. We have seen above that the pleasure of thermal sensation depends upon hypo- or hyperthermia. Hyper- and hypothermia are defined as deviations of deep body temperature from the set temperature, which is approximately 37°C in healthy subjects. In two cases, the set-point is cyclically reset.

In the nycthemeral cycle, the set-point oscillates with a period of 24 h. During this cycle, the pleasure of thermal sensation is precisely adjusted to the defence of the oscillating set-point. As a result of this alliesthesia, behavior is also adapted to the defence of the oscillating set-point; preferred ambient temperature and dressing behavior anticipate the oscillation. Similar oscillations of alliesthesia and dressing behavior occur during the ovarian cycle and defend the 28-day hormonal resetting of the set-point

4.4 Pleasure and Comfort. Thermal comfort was defined by the Association of Heating and Refrigeration Engineers (ASHRAE), as the absence of unpleasant thermal sensation. However, such a definition includes highly unstable states, e.g., a hypothermic subject in a sauna bath. The recognition that sensory pleasure occurs in dynamic states when the environment is useful led the IUPS Commission on Thermal Physiology to propose the following, narrower definition: subjective indifference to the thermal environment [10]; i.e., indifference occurs only in stable states when core temperature is normal and skin temperature neutral.

4.5 Conclusion. This short summary of thermoregulatory behavior and its determinism underlies the close relationship of behavior with thermoregulation. The same signals are responsible for autonomic responses and for sensory pleasure, comfort, and behavior. Thus, it is understandable that the pathology of thermoregulatory behavior will be closely related to that of the thermoregulatory system. In the following, we shall examine situations when behavior puts thermoregulation on jeopardy, then,reciprocally, situations when inadequate thermoregulation entails discomfort and thermoregulatory behavior.

5. Behaviors that Influence or Jeopardize Temperature Regulation

When behavior is devoted to satisfying motivations with higher priorities than temperature regulation, the result may be dangerous to the subject. To satisfy other motivations may entail excessive heat loss, excessive heat gain, or excessive heat production. Other behaviors may also hinder the normal functioning of the central nervous system, especially through drug intoxication.

5.1 Behavior In Health

5.1.1 Exposure to Cold and Warm Environments. If a subject is motivated to expose himself to an extreme environment, the laws of physics apply to his body, and the amount of heat loss or gained depend on the environment. Such a situation is a simple conflict of motivations; the resulting behavior adopted by the subject is that which satisfies the stronger motivation. Extreme cases are exemplified in fire-fighters, in workers having to dive into hot water or to exercise in hot environments, and in space flights. More moderate, but still potentially lethal situations are faced when religion, sport, or military sense of duty are stronger than the motivation for temperature regulation. One way to prolong the tolerance to extreme environment is to acclimatize the subject, e.g., with the hyperthermia of muscular exercise (see below), but as well with exposure to a cold environment. The best way to avoid thermal accidents is to monitor the subjects' thermal discomfort when exposing them to thermal hazards.

B-4

5.1.2 Sleep. Unfavorable environmental temperatures, warm as well as cold, hinder sleep. When a meal before sleep is spicy, the subsequent perturbed sleep may be attributed to effects on temperature regulation. On the other hand, perturbed sleep only slightly affects body temperature.

During sleep, rectal temperature undergoes larger variations than tympanic (and presumably brain) temperature, and the thermoregulatory responses correlate better with the latter than the former: this may have led to the belief that thermoregulation is less accurate during sleep. Actually, the onset of sleep is characterized by a wave of sweating that fades when body temperature drops. This suggests a simple resetting of the thermostat to a lower value. Parmeggiani [11] was the first to observe in animals that temperature regulation was inhibited during paradoxical sleep. In humans, paradoxical sleep is characterized by a complete inhibition of warm, but less of cold thermo-regulatory responses; as a result, rectal temperature is, then, more dependent on ambient temperature [12]. Vasomotor responses seem to be inhibited during hyper-, but not hypothermia.

5.1.3 Acute Muscular Work and Training. As mentioned above, muscular work may be used by a subject as a deliberate substitute for shivering. Reciprocally, heat stress causes fatigue and limits muscular performance [13]. The various physiological responses to exercise correlate with the body temperature rise. Indeed, a high temperature is considered a favourable condition for muscle work. On the other hand, hyperthermia accelerates the onset of anaerobic work, since the mobilisation of heat loss responses competes with the circulatory needs of the working muscles. Thus, in a warm environment, intense musle heat production can become a hazard if the subject ranks the motivation to work or to compete higher than the thermal discomfort aroused [14]. Training contains such a threat, and improves the physical performance. This results in warm acclimation, i.e., improved sweating at similar body temperatures (see chap. 11). In the short term, proximate protection against hyperthermia in warm environments is obtained from pre-exercise cooling and water or fluid intake during the exercise.

5.1.4 Alcohol Consumption. In a warm environment, alcohol intake increases the urinary flow without changing the plasma volume. This is of little concern if fluid is available ad lib. In a cold environment, alcohol intake is followed by drastic responses. The most visible is skin vasodilatation, which may be considered as favorable to preventing peripheral frostbite; but, at the same time, it enhances eventual hypothermia. An interesting hypothesis was presented by Briese and Hernandez [15] who observed that rats treated with ethanol were seeking a cooler environment. They proposed that alcohol triggers an anapyrexia, i.e., a downward resetting of the biological thermostat. Confirmation of such an hypothesis would be skin vasodilatation and sweating after ethanol intake. Another support of anapyretic influences of alcohol in humans can be found in earlier experiments, when subjects immersed in a cold (22°C) or very cold (13 and 15°C) bath shivered less and vasodilated more under alcohol than during control sessions, and, in addition, felt comfortable in the cold water. An anapyretic effect of ethanol would explain why drunks are found hypothermic in winter on public benches of cold countries. The alcohol-induced drop of their set point makes them feel warm, and they fall asleep in thermal comfort although the environment is dangerously cold. Thus, alcohol ingestion might be used as a way to treat peripheral cold injury, provided that the patient is kept in an environment that would prevent drastic, further drops of core temperature.

5.1.5 Voluntary Control of Responses and Mental Work. Mental activity can affect the thermoregulatory responses. This may occur as a by-product of mental activity or as a deliberate goal. Intense concentration on mental calculus interrupts the pain caused by cold water immersion and further prevents the local adaptation to this stimulus. In warm environment, mental calculus inhibits sweating on the arm, but increases it on the trunk.

During yoga and related activities, thermoregulatory responses can be controlled by the subject's will: skin temperature and sweating can thus be raised or diminished for the duration of the sessions. Similarly, hypnosis and suggestion are able to improve comfort or modify the thermoregulatory responses in cold and warm environments.

Biofeedback is a deliberate use of the capacity to control autonomic responses mentally; thus, skin vasomotor tone and skin temperature can be placed under a subject's own control.

Mental activity can, therefore, modulate the autonomic responses; however, the amplitude and duration of this effect remain limited, and the paramount influence of mental activity on temperature regulation remains the control of behavior rather than autonomic responses.

5.1.6 Emotion. Young boys and their coaches before a boxing competition, students during academic examinations characteristically present elevated core temperatures. Extensive animal experimentation by Briese et al. [16] have shown that this elevated core temperature is emotional, because habituation diminishes the rise caused by handling, and that it is a fever (i.e., due to a raised set-point) because it is inhibited by salicylate, and because skin vasomotion defends the higher set-point (see also chap. 10).

5.1.7 Hand Local Cold Adaptation. Repeated cold exposure of the hands entails a local adaptation characterized by intense permanent vasodilatation which prevents the lowering of skin temperature, even in ice-cold water. This local adaptation to cold, which is accompanied by a loss of cold pain, is called cold-induced vasodilatation (CIVD) and is common among individuals who expose their hands continually to cold, e.g., some fishermen.

5.2 Behavior In Disease

5.2.1 Therapeutic Whole-Body Hyperthermia. Whole-body hyperthermia is used in patients suffering from disseminated cancers as a way to potentiate chemo- and radiotherapy. Since these whole-body hyperthermias reach temperatures up to 42°C, i.e., are lethal if prolonged, an excellent knowledge of the physiology of temperature regulation is a prerequisite for this kind of therapy. Psoriasis also was reported to be improved by hot baths.

5.2.2 Anesthesia. General anesthesia abolishes drastically all behavior of the patient. In addition, thermoregulation is also abolished, together with consciousness. Since the patient becomes ectothermic, the physician must take charge of temperature regulation.

This classic view of the influence of general anesthesia was recently challenged [17] when it was observed that anesthetized patients simply broadened their neutral zone of core temperature. Rather than total suppression of temperature regulation, the patients become ectothermic over a zone of up to 4°C, depending on the dose and nature of the anesthetizing agent. It is only between the thresholds for warm and cold defence responses that the patient is ectothermic. Beyond the new thresholds of core temperature, all the thermoregulatory responses are intact, and both core and skin temperatures influence the responses as in non-anesthetized subjects.

5.2.3 Iatrogeny. Some drugs also may interfere with temperature regulation. Such a disturbance may be lethal following drastic hypothermia or malignant hyperthermia. For example, benzodiazepines and certain neuroleptic drugs may produce hypothermia, but the opposite action, malignant hyperthermia, has also been reported in susceptible individuals. Besides therapeutic interventions, innumerable other drugs have been reported also to influence temperature regulation one way or another.

5.2.4 Nervous and Mental Diseases. Catatonic or cataleptic attacks during schizophrenia completely abolish all behavior, including thermoregulatory behavior; there is, however, no study of body temperature and thermoregulation in such patients. A similar absence of thermoregulatory behavior is found in patients suffering from congenital indifference to pain. Although completely normal in all other regards, and in spite of correct qualitative and quantitative thermal sensations, these patients lack the affective dimension of sensation. As a result, they do not feel discomfort and, hence, do no feel the urge to thermoregulate behaviorally. These subjects must be taught how to behave regarding temperature stress. In advanced leprosy, the patients lose their cutaneous thermal sensation. As a result, their thermoregulatory behavior is maladapted and relies only on the perception of discomfort when their core temperature deviates from normal.

Jeddi [18] has shown a pathologic taxis for warm stimuli in psychotic children who were attracted by hot radiators to the point of burning themselves. Pain was shown to suppress alliesthesia aroused in temperature sensation, i.e., when an intense pain was present, thermoregulatory behavior became a lower ranking motivation. Finally, some psychotic patients are especially sensitive to weather changes, including variations in ambient temperature.

6. Modified or Pathological Temperature Regulation that Influences Behavior

Overwhelmed temperature regulation may impair the capacity to behave normally, while an adequate environment improves performance. Because behavioral and autonomic responses are complementary and can be substituted for one another, behavior will be influenced by situations in which the autonomic responses are impaired or are out of control. Thus, the capacity to respond autonomically influences the thermopreferendum. For example, a patient with advanced amyotrophy, who complained permanently of cold discomfort, defended a stable core temperature behaviorally in a broad range of environments. In that patient, the lack of muscular thermogenesis was compensated by behavior. Less anecdotal reports show that paraplegics, who also have diminished muscular thermogenic power, similarly tend to become hypothermic in lukewarm environments and must rely on behavior to thermoregulate.

6.1 Modified Responses. The thyroid gland is a major factor in heat production. Thermophobia in hyperthyroidic patients, and cold discomfort in hypothyroidic patients are well known symptoms. This was confirmed by animal experiments in which thyroidectomized rats sought more heat than controls, while dogs and rats treated with thyroxin sought a cooler environment.

Anhidrosis is a congenital lack of sweat glands. As a result, the patients are heat-intolerant. However, when they adjust their behavior, they are able to live normally and even to participate in competitive sports. The opposite syndrome is hyperhidrosis, but the thermoregulatory impact is limited because excess sweating takes place only on palms and soles.

More serious are the diseases affecting the vasomotor response. When vasomotion is only slightly impaired, the result is only a small change in the subject's preferred ambient temperature. When the vasoconstrictor tone is enhanced pharmacologically, as with ß-blockers, the patients tend to become more easily hyperthermic. Erythrodermia is a dramatic disease leading to death if untreated. The patients' skin is permanently vasodilated, hence the name of the disease, and they are incapable of retaining enough heat to regulate their core temperature even in a tepid environment; it follows that they shiver almost continuously, complain of permanent cold discomfort, and thermoregulate behaviorally.

6.2 Ageing. Considerable work has been devoted to the study of the influence of ageing on temperature regulation (see reviews in [19, 20]), but little conclusive information is available, especially with regard to thermal comfort. As with the other functions, temperature regulation is affected by ageing, and it is not clear whether the changes result from ageing or sedentarism combined with the absence of taxing the responses. It is also not clear whether the low and high core temperatures reported in elderly patients are the result of increased tolerance (i.e., loss of internal sensitivity) or decreased efficacy of the responses.

In a cold environment, thermogenesis and motivation are often diminished in the elderly. However, if basal metabolism and food intake of old men are lower than those of younger controls when expressed in absolute values, the influence of age disappears when expressed as a function of body mass. Interestingly, it has never been hypothesized that the elderly poor, who cannot afford to heat their homes in cold climates, might be acclimatized to cold. This would make them similar to those primitive populations who used to sleep in the open, naked, or to cold-acclimatized individuals who do not feel discomfort when their rectal temperatures drop.

In a warm environment, age does not influence importantly autonomic nor behavioral heat defence. Elderly subjects are reported to maintain behaviorally a greater stability of their skin temperature under their clothes. Age does not diminish the thermoregulatory autonomic response to heat, but limits the cardiac response to heat stress. The number of active sweat glands tends to diminish, even in physically fit subjects, but the capacity to sweat is not impaired. The fact that the old tend to sweat and vasodilate more heavily on their foreheads allows hypothesizing that they have an improved and more efficacious selective brain cooling. If this is the case, one would expect their trunk temperature to be higher; this indeed has been reported. Thus, hyperthermia in elderly subjects might be simply the open-loop result of selective brain cooling on trunk temperature [21].

B-9

6.3 Sexual Differences. The difference of attitude between men and women regarding ambient temperature is another well-documented debate. Thus, it is often stated that women report both cold and warm discomfort earlier than men. However, this is controversial.

In warm environments, women are generally found to vasodilate and sweat later than men. As a result, their skin and rectal temperatures are higher. However, before accepting a sexual difference in warm defence, several remarks must be made. Part of the difference may be due to the resetting of the female thermostat at a 0.3-0.5°C higher set-point during the luteal phase (see below). The higher set-point found in women may be explained by the lumping of all female data without regard to the period of the cycle. Several studies have shown that, indeed, women sweat less than men and have higher skin temperatures, but the resulting rectal temperature is equal in both sexes; physically fit or warm-acclimatized women sweated more than untrained men. In one study of non-acclimatized subjects, women were even found to have a better thermolytic response than men. Rather than a higher intolerance to heat, women reported discomfort to occur at higher body temperatures. This, together with the delayed evaporative response often reported in women, should be studied in the light of the recently demonstrated selective brain cooling. Similar to the hypothesis offered above for the elderly, it cannot be disregarded that the females' delayed response to heat could be also the result of better selective brain cooling.

In cold environments, women complain of cold discomfort much earlier than men. Especially, they complain more of cold feet discomfort, but this difference might be related to the different clothing worn by the two sexes.

In addition, such a difference is not likely to be of significance because, reciprocally, men are more susceptible to cold stress than women when performing mild exercise. Indeed, no difference between the sexes is found when the experiments are carefully designed.

In conclusion, the sexual differences found in the literature regarding responses to cold as well as to warm stress, and the perception of thermal comfort seem to be limited to the changing set-point of the menstrual cycle (see below), i.e., sexual differences, if they exist, are much smaller than individual differences.

6.4 **Modified Set-Point.** We saw above that the set-point for temperature regulation is an important signal for the perception of thermal comfort/discomfort. Shifts in set-point result in thermal discomfort and changes of thermal preference (Fig. 2). This was studied in various physiological and pathological circumstances.

During infectious fever, the set-point for temperature regulation is raised. This entails autonomic responses to defend the new set-point; in addition, the seeking of thermal comfort also serves to defend the raised set-point.

Anapyrexia is the reverse of fever. In this syndrome, the set-point is lowered, and the subjects use autonomic and behavioral responses to defend the low set-point. Periodic anapyrexia has been described in the literature, sometimes linked for unknown reason to an agenesis of the corpus callosum. Menopausal hot flushes are an example of anapyrexia. During the flush, a seizure-like anapyrexia occurs with all the autonomic responses activated to the defence of the low set-point. At the same time, warm discomfort triggers a behavior also adapted to the defence of the lower set-point.

The ovarian cycle is characterized by a core temperature lower in the follicular phase than in the luteal phase of the cycle. A study of thermal pleasure showed that this change is an adjustment of the set-point; the seeking of pleasure triggered a behavior defending the adjustable set-point. The concomitant increase in the threshold temperature for all autonomic thermoregulatory responses supports the concept of a resetting of the set-point. The fact that the amplitude of the autonomic responses is the same in the follicular and luteal phases of the cycle would confirm that the only variable of temperature regulation affected by the ovarian hormones is the set-point.

The nycthemeral cycle is another well-known example of adjustable set-point. The set-point oscillates with a period of 24 h, and behavior defends the oscillating set-point.

Paradoxical shifts of the set-point are described in the late stages of hyperthermia and hypothermia. When temperature regulation is overwhelmed, suddenly the thermostat shifs drastically towards higher values in heat stroke, and lower values in hypothermia. In both cases, behavioral responses defend the new set-points. The result of such resettings on the border of death is the abandonment of any autonomic response in subjects whose physiological capacity is already overtaxed. These symptoms are bad prognoses as they are the last defences of dying patients.

Finally, a higher resetting of the temperature set-point is reported in anorexia nervosa, while the opposite trend is reported in obesity. This is evidence that the systems controlling thermoregulatory behavior and food intake may overlap in the brain.

7. Conclusions

Temperature regulation is so important for survival that multiple, redundant, regulatory loops operate simultaneously, with the result that core temperature is well defended. Two main factors may impair thermoregulatory behavior, viz., pathology and the subject's own will.

A physician who knows rather little about temperature regulation should be prudent with the patients' thermal diseases, and conservative when the patients are able to behave. The overall lesson that emerges from this short review is that pathology is rich but rarely dangerous, except in rare, extreme cases when behavior is suppressed or in psychotic patients. Behavior is the most efficacious response an individual can muster to thermoregulate; it is more powerful and can be sustained longer than the complementary autonomic responses. Thermal comfort is in tune with thermoregulation, as similar signals trigger autonomic and behavioral responses. The physician should, therefore, look to the patients' thermal discomfort as a useful symptom for diagnosis and appropriate therapy.

The subject's own will is far more dangerous. Because behavior is a final common path, various motivations must share it to satisfy competing goals. When a subject considers that his sport, sense of duty, or religious motivation ranks higher than his physiological welfare, the lethal threshold of heat stroke or of hypothermia may be reached.

8. References

1. Fanger PO (1970) *Thermal Comfort*. København: Danish Technical Press. 244.
2. Cabanac M, Brinnel H (1987) The pathology of human temperature regulation: Thermiatrics. *Experientia*; **43**:19-27.
3. Cabanac M, Massonnet B, Belaiche R (1972) Preferred hand temperature as a function of internal and mean skin temperatures. *Journal of applied Physiology*, **33**:699-703.
4. Hardy JD (1965) The 'set-point' concept in physiological temperature regulation, in *Physiological Controls and Regulations*, Yamamoto WS, Brobeck JR, Editors. Saunders and Co.: Philadelphia, 98-116.
5. Mrosovsky N (1990) *Rheostasis, the Physiology of Change*. New York: Oxford University Press. 183.
6. Cabanac M. (1981) Physiological signals for thermal comfort, in *Bioengineering, Thermal Physiology, and Comfort*, Cena K, Clark JA, Editors. Elsevier Scientific Publ.: Amsterdam, 181-192.
7. Bleichert A, Behling K, Scarperi M, Scarperi S (1973) Thermoregulatory behavior of man during rest and exercise. *Pflügers Archiv*, **338**:303-312.
8. Marcus P, Belyavin A (1978) Thermal sensation during experimental hypothermia. *Physiology and Behavior*, **21**:909-914.
9. Attia M, Engel P (1982) Thermal pleasantness sensation: an indicator of thermal stress. *European Journal of applied Physiology*, **50**:55-70.
10. IUPS Commission for Thermal Physiology (1987) Glossary of terms for thermal physiology. *Pflüg ers Archiv*; **410**:567-587.
11. Parmeggiani PL (1970) Sleep and environmental temperature. *Archives italiennes de Biologie*, **108**:369:387.
12. Shapiro CM, Moore AT, Mitchell D, Yokdaiken M (1974) How well does man thermoregulate during sleep. *Experientia*; **30**:1279-1280.
13. Nielsen B. (1992) Heat stress causes fatigue! in *Medicine and Sport Science*, Hebbelinck H, Shephard RJ, Editor. Karger: Basel, 207-217.
14. Wyndham CH (1973) The physiology of exercise under heat stress. *Ann Rev Physiol*; **35**:193-220.
15. Briese E, Hernandez L (1996) Ethanol anapyrexia in rats. *Pharmacology Biochemistry and Behavior*, **54**:399-402.
16. Briese E, de Quijada M (1970) Colonic temperature of rats during handling. *Acta physiologica latinoamericana*, **20**:97-102.
17. Sessler DI (1993) Perianesthetic thermoregulation and heat balance in humans. *FASEB J*; **7**:638-644.

18.   Jeddi E (1970) Confort du contact et thermorégulation comportementale. *Physiology and Behavior*, 5:1487-1493.
19.   Khogali M, Hales JRS, ed. (1983) *Heat Stroke and Temperature Regulation*, Academic Press: Sydney.
20.   Shiraki K, Yousef M, ed. (1987) *Man in Stressful Environments. Thermal Work and Physiology*, Ch. Thomas: Springfield.
21.   Cabanac M (1995) *Human Selective Brain Cooling*. Neuroscience Intelligence Unit, Austin TX: R.G. Landes Co & Springer Verlag. 114.

Self-evaluation

-List the various categories of thermoregulatory behaviors and provide examples for each category.
-Give examples of thermoregulatory equations describing autonomic and behavioral responses.
-Give the definition of thermal comfort.
-What are the physiological signals for thermal comfort?
-List some motivations that can jeopardize temperature regulation.
-What is the influence of sleep on temperature regulation?
-What is the influence of muscular work on temperature regulation?
-What is the influence of alcohol consumption on temperature regulation?
-Is mental activity capable of modifying temperature regulation?
-What is emotional fever?
-What is whole-body hyperthermia?
-What is the influence of general anesthesia on temperature regulation?
-Give some examples in which disease modifies thermoregulatory responses; what becomes of behavior in each case?
-What is the influence of ageing on temperature regulation?
-Are there sexual differences in temperature regulation? And in the perception of comfort?
-What is the influence of fever on temperature regulation? And on the perception of comfort?

Chapter 8

Learning Objectives
1. To understand the effects of exercise on thermoregulation in humans.
2. To be able to describe the limitations to physical performance as influenced by circulatory capacity and maximal sweat loss and to environmental thermal conditions.
3. To name non-thermal factors which interfere with and influence thermoregulation during exercise.

Bullets
B-1   The efficiency of external work is less than 20–25%.
B-2   Core temperature is increased during exercise in proportion to work intensity.
B-3   Core temperature during exercise rises in proportion to the relative workload of the individual.
B-4   Mean skin surface temperature depends on ambient conditions independently of heat production.
B-5   In the "prescriptive zone", body core temperature during exercise is uninfluenced by environmental temperatures.
B-6   Maximal sweating rate is important for performance in hot conditions.
B-7   Competition between muscle and skin blood flows causes an increase in core temperature in hot conditions.
B-8   The fatter individual tolerates cold better.
B-9   The process of heat acclimation leads to increased sweating and larger blood volume.
B-10 The thermoneutral zone is reduced in water.
B-11 Sweat loss impairs exercise performance.
B-12 Besides the % $\dot{V}O_2/T_{core}$ relationship, physical fitness and inactivity have specific effects on exercise core temperature.
B-13 Substrate utilization may affect temperature regulation.
B-14 Sex hormones may interfere with thermoregulation in women.
B-15 Children and the elderly have reduced thermoregulatory capacities.
B-16 Is it true that warming-up is beneficial for performance?

Chapter 8

TEMPERATURE REGULATION IN EXERCISE

BODIL NIELSEN

August Krogh Institutet, University of Copenhagen, Denmark

AND

HANNA KACIUBA-USCILKO

Dept. Applied Physiology, MRC, Polish Academy of Sciences, Warsaw, Poland

8.1. THERMAL FACTORS.

8.1.1. Introduction.

During exercise, the rate of heat liberation is increased due to the increased metabolic energy turnover in the exercising muscles. The efficiency of these processes in converting the chemical energy to external work is less than 20--25 %.

The efficiency, E, is defined as

$$E\% = \frac{\text{external work} \cdot 100\ \%}{\text{metabolic energy cost}} \quad (1)$$

Therefore the remaining 75--100% of the liberated energy appears as heat in the active muscle tissue.

The temperature in the muscles rises rapidly at the onset of exercise, within 3--6 min from about 34°C during rest to 37--38°C in the large muscles of the thigh during bicycling. The heat liberated in the muscles is conducted to the blood flowing through them, and in this way distributed to the rest of the body core, promoted by the increased circulation to the exercising muscles.

8.1.2. The Change in Core Temperature.

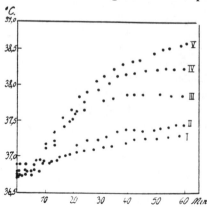

Fig. 1: Rectal temperature during work (subj. P.J.), at work intensities I: 60W, II: 90W, III: 150W, IV: 180W and V: 210W. From [1] with permission.

Core temperature can be measured with thermocouples inserted into the rectum (rectal temperature, T_{re}) or into the deep esophagus (esophageal temperature, T_{es}). The latter responds faster to exercise. Esophageal temperature near the cardia is an index of the temperature of the blood in the heart and aorta, due to the vicinity of the esophagus to these. At the onset of exercise, the core temperature starts to rise, and after 30--40 min of sustained exercise, the core temperature plateaus; this increased level compared to resting values is maintained as long as the exercise is continued [1]. The rise is greater the higher the work load. This means that heat is initially stored in the body, and first after a certain increase in temperature the heat liberation is balanced by a dissipation of heat at the same rate. This is when the temperature plateau is reached. Fig. 1 illustrates the increase in body core temperature during exercise of increasing intensities.

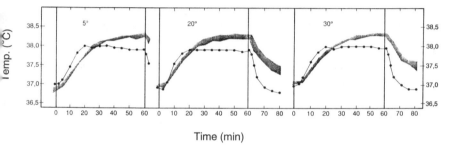

Fig. 2: Rectal temperature at four depths (12--27 cm, shaded area) and deep esophageal temperature (filled circle) during 60 min bicycling exercise of 150W. Environmental temperatures 5°C, 20°C and 30°C. Average results from 4 exp. in each condition in one subject. Adapted from [2].

The increase in core temperature appears to be regulated in proportion to the intensity of exercise. This is indicated by the fact that the exercise core temperature is independent of the environmental temperature and, hence, the conditions for heat loss. Fig. 2 illustrates this; the same exercise intensity (150W) was performed at 5°, 20° and 30°C environmental temperatures [2]. In spite of the different conditions for heat loss the core temperature reached the same level in all conditions.

B-2

Furthermore, the exercise core temperature is determined by the relative work load, i.e., the percentages of the aerobic capacity, $\dot{V}O_{2max}$ of the individual [3].

B-3

When different individuals exercising at the same environmental temperature and the same work intensity and oxygen consumption are compared, their core temperature at steady state will stabilize at different levels, in spite of the fact that the rate of heat production is the same (Fig.3 left). However, at the same relative load, e.g., at 50% of $\dot{V}O_{2max}$ the rise in core temperature is similar in all individuals; it increases 1°C, e.g., from 37°C during rest to a steady state of 38°C. In other words, the exercise core temperature is independent of the rate of heat production, heat dissipation and absolute oxygen consumption. (Fig.3, right).

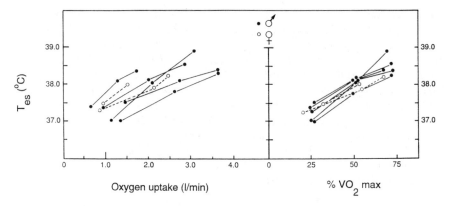

Fig. 3: Esophageal temperature in steady state after 60 min exercise in 7 different subjects. Plotted against absolute oxygen uptake and work load (left) and against relative oxygen uptake, % of max $\dot{V}O_2$ (right). Adapted from [3].

$\dot{V}O_{2max}$ is determined by the maximal capacity of the circulatory system for transport of oxygen, illustrated in the Fick equation:

$$\dot{Q} = \frac{\dot{V}O_2}{(a\text{--}v)\ O_2\ \text{diff.}} \qquad (2)$$

or $\qquad \dot{V}O_{2max} = \text{max HR} \cdot \text{max SV} \cdot \text{max (a--v) } O_2 \text{ diff.} \qquad (3)$

$\qquad \dot{Q} = \text{heart rate} \times \text{stroke volume} \qquad (4)$

where $\dot{Q}$ = Cardiac output, $l \cdot min^{-1}$
$\dot{V}O_2$ = Oxygen uptake, $ml \cdot min^{-1}$
a = Arterial oxygen content $ml \cdot l^{-1}$
v = mixed venous oxygen content, $ml \cdot l^{-1}$
HR = heart rate, $beats \cdot min^{-1}$
SV = stroke volume of the heart, ml

The theoretical maximal O_2 uptake occurs during exercise when the heart rate, HR, is maximal. Maximal HR is dependent on age (approx.(220--age) beats per min). Stroke volume depends primarily on body size, but is increased by endurance training.

It is still not known how the increase in body core temperature is regulated during exercise in accordance with the maximal aerobic capacity. Signals could arise from metabolic or hormonal factors liberated in proportion to the relative severity of the load on the individual. However, the rise in core temperature acts as the main stimulus in the control of sweating and skin circulation.

8.1.3. Skin Surface Temperature during Exercise.

The temperature of the naked skin surface is not uniform; more distal parts (e.g., hands, arms, feet) are cooler than the skin of the head and the trunk. It is possible to calculate a mean skin temperature, T_{sk}, by which the temperature of the different skin areas are "weighted" in proportion to the fraction of the total body surface they represent (see also Chap.2). The conditions for heat exchange depend on the temperature difference between the body surface and the environmental temperature for the convective and radiative heat exchanges, and on the amount of sweat which can be produced and evaporated (see Chap.3).

The mean skin temperature is independent of the exercise intensity and heat production within the prescriptive zone (see below). Thus, at a certain environmental temperature, the heat loss by convection and radiation is the same, while, with exercise at higher intensities and heat production, the sweating will increase in correspondence to the difference between the heat production and the heat lost by convection and radiation. With increasing environmental temperature, the T_{sk} increases, and the difference between it and the environmental temperature, T_a, decreases. At approx. 35°C, the T_{sk} equals T_a. At even higher T_a, the gradient is reversed. When the environmental temperatures (air and mean radiant temperatures) are higher than the skin surface, heat is gained from the environment by convection and radiation. The only route for heat loss under such conditions is by evaporation of sweat.

B-4

8.1.4. The Prescriptive Zone.

The body core temperature during exercise is uninfluenced by environmental temperature over a wide range of temperatures, due to the thermoregulatory control of the heat loss mechanisms, viz., skin blood flow and sweating. The higher the skin temperature, and thus the smaller the possible convective and radiative heat loss, the more sweating is elicited. The interactions of stimuli in this control are described in Chap..

The range of environmental conditions in which the body temperature is independent of these conditions is called the prescriptive zone [4]. Above the <u>upper critical temperature</u>, core temperature increases to a higher level than during exercise in thermoneutral conditions, while on the other hand, below the <u>lower critical temperature</u> the core temperature falls. The actual range of the prescriptive zone, also called thermoneutral zone, depends on the rate of heat production and, furthermore, on the physiological capacity for heat dissipation for the upper limit, and on the maximal rate of heat production and vasoconstriction for the lower critical temperature (Fig.4).

B-5

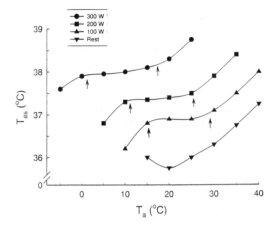

Fig. 4: The esophageal temperature after 2 hr of exercise in one subject working at different exercise intensities (100--300W) at environ mental temperatures between --5°C and 40°C. Note how both upper and lower critical temperatures (marked by arrows) move to the left with increasing work load and heat production. (Adapted from Kitzing et al; Int. Z. angew. Physiol. 30: 119-131, 1972).

The physiological capacity for heat dissipation is linked to the ability to sweat. This depends on the size of the individual, on the physical fitness, and on the state of heat acclimatization. Maximal sweating rates may vary between 600--700 ml h^{-1} for a sedentary person, to about 4 l h^{-1} in very well trained and heat acclimated individuals exercising in dry heat (see Chap.11.1).
The circulatory capacity also affects the ability to sustain exercise in the heat and, hence, the upper critical temperature.

The amount of blood needed for the transport of heat to the skin, H_{skin}, is expressed by the equation:

$$H_{skin} = \dot{Q}_{sk} \cdot c \cdot (T_{ar} - T_v) \qquad (5)$$

where $\dot{Q}_{sk}$ is blood flow to the skin in l·min, c the heat capacity of blood (approx·4kJ·kg^{-1}) and T_{ar}--T_v is the temperature of arterial and venous blood, respectively, reaching and leaving the skin. If we substitute T_{ar} with T_{core}, rectal or esophageal and T_v with T_{sk}, and rearrange the equation, we obtain

$$\dot{Q}_{sk} = \frac{H_{skin}}{c \, (T_{re} - T_{sk})} \qquad (6)$$

In warm conditions the difference (T_{re}--T_{sk}) becomes smaller and the skin blood flow necessary to carry the heat to the skin increases (see Eq.5), until a limit when the ability of the heart both to supply blood to the exercising muscles and to cover the thermoregulatory demand for skin blood flow is reached. Under this condition, the core temperature increases, and the skin blood flow may be reduced (cf. Eq.6). This is when the upper critical temperature is surpassed.

The lower critical temperature is determined by the balance between external heat loss in conditions of maximal insulation, when maximal vasoconstriction is elicited, as well as heat production by shivering and exercise [5]. A fatter individual can sustain lower environmental temperatures due to the greater insulation of the peripheral tissue layers. This, on the other hand, is a disadvantage in the heat, not being able to reduce this insulation so much as a leaner individual, albeit blood flow to the skin is not impeded. Again, acclimation to cold (Chap.11.2) may increase the tolerance to cold exposure by increasing both shivering and non--shivering thermogenesis, as well as vasoconstriction, the latter increasing the insulation of the peripheral tissues.

B-8

8.1.5. Acclimatization to Heat

Stressful environments induce physiological adaptative changes [5], which improve the tolerance to the stress (see Chap.11.1). When humans are exposed acutely to exercise in hot environments, their heart rate and core temperature increase more than under cool conditions, and their performance and endurance for prolonged exercise are reduced. With repeated daily exposures over a period of several days to weeks, physiological changes occur which include an increase in sweating rate, a lowering of resting core temperature, and increased blood volume. These adaptations lead to an increased evaporative heat loss, resulting in the lowering of the core and skin temperatures during work under hot conditions. Furthermore, the improved filling of the cardiovascular system results in a lower heart rate and improved endurance for exercise.

B-9

8.1.6. Exercise in Water.

The thermal properties of water differ from those of air by a 20--fold higher heat conductivity and a specific heat about 1000 times greater than air. The average skin temperature in a water environment will, therefore, be very close to the water temperature; e.g., for persons swimming in water, it is 0.2--0.75°C higher than the water temperature. In warm water, the physiological mechanisms for varying the heat loss depend largely on the mechanisms for changing the heat flow, i.e., the blood flow, to the skin. The heat transfer through the skin also varies with the amount of fat in the subcutaneous tissues both in warm and cold water. The heat transfer by the blood can vary 6--10 times above a minimal value, and the blood flow can provide a maximal heat transfer of about 100 W m^{-2} °C^{-1}. From this it can be calculated that a subject swimming with a rate of oxygen consumption of 3 l min^{-1} cannot maintain a steady body temperature if the water is warmer than 32°C, which is the upper critical temperature in this condition [6].

B-10

In cold water the skin blood flow is reduced to a minimum. The minimum heat loss under these circumstances is related to the fat layers in the subcutis. During rest, a fat subject (36% of body weight fat) can maintain heat balance in water 10°C lower than core temperature, while a lean person (less than 10% of body weight in fat) can tolerate a water temperature no more than 3.8°C lower than his body temperature without shivering during rest. In other words, the prescriptive zone in water is very narrow, -- only about 5°C.

The next line of defence against cooling, after the blood flow has been minimized, is shivering. Shivering may increase heat production by approx. 500W, and this shivering heat is added to the metabolic heat production of the ongoing activity in the defence of body temperature (see Chap.3).

8.2. NON--THERMAL FACTORS.

8.2.1. Introduction

Thermoregulatory responses to exercise depend not only on thermal conditions under which physical work is performed, but also on many non--thermal factors. Among these are, e.g., the state of body hydration, the blood volume, the ionic composition of body fluids, as well as the availability of energy substrates and the fitness of the subjects performing exercise [7]. Besides, in women, thermoregulatory responses to exercise and, in particular, the onset and magnitude of sweating, depend on their menstrual phase. Furthermore, thermoregulation during exercise changes with age, an important factor to be considered (see Chap. 9.2.).

Excessive increases in core and muscle temperatures markedly reduce endurance performance. Thus, some procedures purported to improve effectiveness of thermoregulation, such as active warming--up or precooling maneuvers, have been introduced before starting to exercise; the mechanisms of their effects are relatively well understood.

8.2.2. State of Hydration

During exercise, water and electrolytes are lost due to sweating. Sweat is usually hypotonic; therefore, sweat loss is accompanied by increased osmotic pressure of extracellular fluids since body electrolytes become more concentrated. Even relatively small declines in body water content (1% of body mass) can impair exercise thermoregulation and lead to a decrease of work tolerance, particularly during long--term activity [8,9]. Water is lost from all body water compartments. The reduction of blood volume, due to this dehydration, decreases cardiac output and blood pressure, which in turn diminishes blood flow to the muscles and skin. Because less blood reaches the skin, heat dissipation is attenuated, i.e., the body retains more heat and, hence, the core temperature increases (see Eq.6). In long-

-distance runners, sweat losses can approach 6% to 10% of body mass. Such severe dehydration can limit subsequent sweating, and make an individual susceptible to heat--related illnesses [10]. Besides the cardiovascular effects of hypohydration, the impaired function of the thermoregulatory mechanisms during physical work under these conditions could be caused by the effects of cellular dehydration on the hypothalamic neurones connected with body temperature regulation. The exercise--induced dehydration can be reduced by drinking water prior to or during exercise [11,12]. This, to some degree, limits the excessive increase in body temperature in exercise by shortening the delay in onset of sweating and decreasing the amount of dripping sweat (Fig.5). As a result, work performance can be markedly extended.

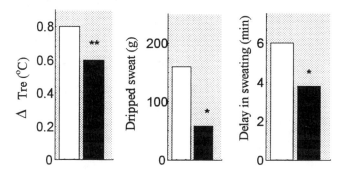

Fig. 5: Effect of hyperhydration on rectal temperature (Tre) increment, amount of dripped sweat and delay in sweating in men exercising for 45 min at 50% of $\dot{V}O_{2max}$. White bars - euhydrated subjects (normal water intake), black bars - hyperhydrated subjects (intake of 2 l of tepid water during 1 h prior to exercise). Adapted from [12].

8.2.3. Ionic Composition of Body Fluids

Equilibrium levels of rectal temperature during exercise in humans are highly correlated ($r \geq 0.71$) with the plasma sodium and osmolar concentrations. The ionic and osmolar factors appear to act by controlling sweat gland function because the rate of sweating is inversely proportional to the plasma ionic and/or osmolar concentration. These factors also modify body temperature responses to exercise in dogs that dissipate most of their heat by panting. In this species, an excessive increase in core temperature can be induced by extracellular hyperosmolality per se without elevating sodium concentration in the extracellular fluid. Furthermore, data obtained both in animal experiments and in humans suggest that the magnitude of exercise--induced increase in core temperature may be brought about by a decreased concentration of calcium ions in blood and cerebrospinal fluid.

8.2.4. Physical Fitness

In spite of the general correlation between $\dot{V}O_{2max}$ and core temperature during exercise, human subjects with different work capacities ($\dot{V}O_{2max}$) also differ in thermoregulatory responses to prolonged exercise even when it is performed at the same relative load (expressed as percentage of $\dot{V}O_{2max}$) and under identical thermal conditions [13] (Fig.6). At high work intensities (65% $\dot{V}O_{2max}$), sedentary men achieve approx. 0.8°C higher rectal temperatures than fit subjects, although the former performe lighter work from the point of view of energy cost (Fig.6). These findings suggest that physically trained persons possess a considerably improved efficiency of thermoregulation during exercise, mainly due to enhanced heat loss concsequent to increased skin blood flow and sweating rate. Studies on the time course of thermoregulatory changes during a 3--months endurance training in a temperate envinronment in sedentary men [14] showed that this improved thermoregulatory function during training was not directly related to an increment of aerobic capacity. Thus, during the last two months of training, the faster activation of sweating, that resulted in smallere elevations of the auditory canal temperature in response to exercise (at 50% $\dot{V}O_{2max}$), was not associated with any substantial enhancement in VO_{2max} (Fig.7).

Contrary to increased physical activity, prolonged bed rest (12--14 days), head--out thermoneutral water immersion (6 h) in humans, as well as cage confinement (4--6 weeks) in dogs result in consistent, excessive elevations in core and muscle temperatures (the latter measured only in dogs) during subsequent exercise [15]. The increased temperature during exercise in humans after prolonged bed

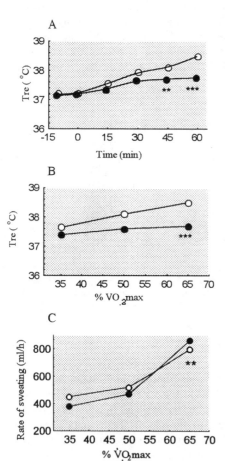

Fig. 6:   (A) Changes in rectal temperature (Tre) during 1 h exercise at 65% $\dot{V}O_{2max}$. (B) and (C) relationships between percentage of (% VO_{2max}) and Tre or rate of sweating. Sedentary subjects - open circles, well fit subjects - filled circles. Adapted from [13].

rest--deconditioning appears to be caused by attenuation of conductive heat dissipation due to reduced peripheral vasodilation. The enhanced exercise--induced increase in core temperature after prolonged restriction of physical activity in dogs is attributed mainly to the reduced respiratory evaporative water loss.

8.2.5. Availability of Energy Substrates for Muscle Contraction

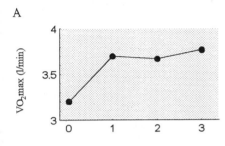

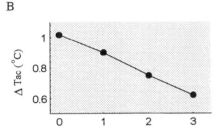

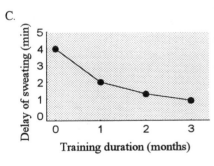

Fig. 7: Effect of endurance training on: (A) aerobic capacity ($\dot{V}O_{2max}$), (B) an increase in auditory canal temperature (Tac) during 1h exercise at 50% $\dot{V}O_{2max}$. (C) delay of sweating response. Data compared to resting values at the beginning of exercise. Adapted from [14].

Metabolic responses to prolonged physical effort depend largely on dietary modifications that alter the proportion of carbohydrates and lipids as energy sources for muscle contraction. Unfortunately, little is known about the influence of such modifications on exercise thermoregulation in humans. However, in exercising dogs, attenuation of glucose metabolism and utilization, e.g., by insulin--induced hypoglycemia, results in an excessive increase in rectal temperature. On the other hand, infusion of glucose into both normal and cage--confined dogs markedly diminished the exercise--induced elevation in rectal and active muscle temperatures. Reduced heat production and/or increased heat dissipation have been suggested as possible mechanisms responsible for the beneficial effects of increased glucose availability on exercise thermoregulation. Enhanced utilization of free fatty acids (FFA) due to fat--rich diets results in accelerated increases of rectal and working muscle temperatures in dogs. This is most probably caused by enhanced storage of heat in the body in spite of marked activation of respiratory heat loss [16].

The link between the energy substrates that are combusted during physical exercise and thermoregulatory responses to work remains unknown. One may only speculate that changes in the intramuscular milieu, connected to different avenues of energy production, influence neural inputs from muscle chemoreceptors to the systems controlling various effectors of thermoregulation. It cannot be excluded, however, that modification of some neurohormonal factors, caused by the presence of different energy substrates, may act directly on the thermoregulatory center and/or effectors, e.g., skin blood flow.

8.2.6. Exercise Thermoregulation and Menstrual Cycle in Women

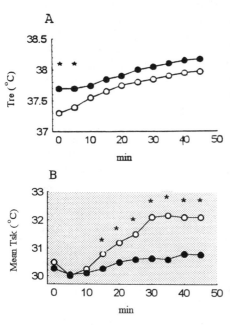

Fig. 8: Effect of menstrual phase on rectal temperature (A) and mean skin temperature (B) in women exercising during 45 min at 50% of $\dot{V}O_{2max}$. Luteal phase - open circles. Asterisks denote significan differences between phases. Adapted from [17].

It has been well established that body temperature fluctuates during the menstrual cycle, but there are only few data on an influence of these fluctuations on thermoregulatory reactions to endogenous or exogenous heat loads; the problem is still a matter of debate. Some studies [17] show that core temperature is exaggerated and skin temperature reduced in women exercising during the luteal phase (L) in comparison with the follicular (F) phase (Fig.8), which again is associated with a higher rate of sweating in L phase. Oral contraceptives, in which progesterone changes mimic natural hormonal changes in the cycle, do not alter the greater in L than in F temperature threshold for sweating, but reduce the phase--related differences in sweating dynamics, thus making the thermoregulatory responses to exercise more uniform during the menstrual cycle.

8.2.7. Changes in Exercise Thermoregulation with Age

Reduced efficiency of thermoregulation in children is one of the factors limiting their ability for endurance exercise. It is partly attributed to the larger ratio of body surface area to body mass in children than in adults, which facilitates heat gain at high ambient temperatures, and heat loss in cold environments. In addition, children up until puberty, have a distinctly deficient sweating ability [18]. Such a deficiency is caused not only by fewer active sweat glands than in adults but also by lower production of sweat per gland (approx. 2.5 times less in comparison with adults). In the post--pubertal period, sweat gland function improves markedly in boys, which results in higher exercise tolerance. In girls, sexual maturation does not seem to exert any significant influence on sweat secretion.

Ageing is often associated with decreased physiological functions, including the ability to regulate body temperature efficiently. The question arises, however, whether chronologic age per se causes poor heat tolerance or whether other factors changing concomitanly with advancing age, e.g., decreased working capacity, sedentary life style, increased prevalence of chronic diseases, etc., play a larger role than age itself (see also Chap.9.2). Thermoregulatory responses to exercise and exogenous heat load in healthy, fit elderly subjects have been compared with those in young ones of similar physiological characteristics [19]. It appears that maintaining a high degree of fitness does not prevent reduction of skin blood flow with age (approx. 25 to 40%) in elderly athletes. Besides, a diminished sweating response to exercise was found in old subjects. In spite of some impairment in these two basic mechanisms of heat dissipation there were no substantial differences in the magnitude of exercise--induced body temperature changes and heat storage between young and elderly athletes.

B-15

8.2.8. Exercise Thermoregulation after Warming--up and Precooling Procedures

Active warming--up has been generally accepted as beneficial for heavy work performance, especially in endurance sports which require rapid adjustment of cardio--respiratory function and metabolism for attaining a state of readiness to exercise. After active warming--up, the onset of sweating during the maximal exercise occurs earlier, while the increases in skin and core temperatures are attenuated in comparison with an identical exercise without warming--up. An important effect of warming--up is the elevation of skin temperature before the start of the subsequent effort, which causes the rise in sweating and, thus, attenuates the rate of increase in exercise core temperature.

B-16

On the other hand, lowering deep body temperature even by 1°C below normal resting value before starting an exhaustive exercise attenuates the exercise--induced increases in core as well as in mean body temperatures at work loads from 20 to 100% of $\dot{V}O_{2max}$. The lower body temperature achieved by precooling does not have any obvious reducing effect on exercise performance. On the contrary, both at submaximal and maximal work rates, the heart rate is significantly reduced, accumulated sweat secretion becomes smaller in comparison with control exercise tests, and $\dot{V}O_2$ is not affected. These effects are probably due to decreased skin and core temperatures after precooling, which result in greater heat dissipation by convection and radiation, and decreased thermal strain and discomfort. It should be emphasised, however, that low pre--exercise temperatures are not recommended for athletes performing short--lasting effort that either requires maximal force and aerobic power from the beginning of exercise or that is performed largely anaerobically, e.g., sprint running.

3. SUMMARY:

Within the prescriptive zone body core temperature is elevated in proportion to the relative load, i.e., to the % $\dot{V}O_{2max}$ of the individual. The capacity for sweating and skin circulation determines the ability for exercise under hot conditions. The efficiency of these effectors may be improved by training and heat acclimatization. Many non--thermal factors influence thermoregulation during exercise. These may influence the hypothalamic thermoregulatory centers, or act peripherally through reflexes from exercising muscles and by direct, local effect on sweat glands and skin arterioles.

References

1. Nielsen M (1938). Die Regulation der Körpertemperatur bei Muskelarbeit. *Skand. Arch. Physiol.* **79**:193-230.
2. Nielsen B, Nielsen M (1962). Body temperature during work at different environmental temperatures. *Acta physiol scand* **56**:120-129.
3. Saltin B, Hermannsen L (1966). Esophageal, rectal and muscle temperature during exercise. *J Appl Physiol* **21**:1757-1762.
4. Lind AR (1963). A physiological criterion for setting thermal limits for everyday work. *J Appl Physiol.* **18**:51-56.
5. Pandolf KB, Sawka MN, Gonzalez RR (eds) (1988). *Human performance physiology and environmental medicine at terrestrial extremes*. Brown & Benchmark, Dubuque, IA.
6. Nielsen B (1978). Physiology of thermoregulation during swimming. In: Eriksson B, Furberg B, Nelson RC, Morehouse CA (eds), *Swimming medicine IV. Internat series on sports sciences, 6.* University Park Press, Baltimore, pp. 297-304.
7. Kaciuba-Uscilko H, Kruk B (1989). Interrelationships between body temperature and metabolism during physical exercise. In: Mercer JB (ed), *Excerpta Medica, Thermal Physiology* 1989, pp. 13-22.
8. Sawka MN (1992). Physiological consequences of hypohydration: exercise performance and thermoregulation. *Medicine & Science in Sports & Exercise* **24**:657-670.
9. Nielsen B (1994). Fluid loss and performance. Insider 2: 1-4.
10. Wilmore JH, Costill DL (1994). Physiology of Sport and Exercise, *Human Kinetics*, Champaign, IL.
11. Greenleaf JE (1979). Hyperthermia and Exercise. In: Robertshaw D (ed) *International Review of Physiology. Environmental Physiol. III*, University Park Press, Baltimore, pp. 157-208.
12. Grucza R, Szczypaczewska M, Kozlowski S (1987). Thermoregulation in hyperhydrated men during physical exercise. *Eur J Appl Physiol.* **56**: 603-607.
13. Kozlowski S, Domaniecki J (1972). Thermoregulation during physical exercise in men of different working capacity. *Acta Physiol. Pol.* 23: 761-772.
14. Smorawinki J, Grucza R (1994). Effect of endurance training on thermoregulatory reactions to dynamic exercise in men. *Biol. Sport* **11**:143-149.
15. Greenleaf JE (1997). Exercise thermoregulation with bed rest, immersion, and confinement deconditioning. *Ann. N.Y. Acad. Sci.* **813**: 741-750.
16. Kaciuba-Uscilko H, Falecka-Wieczorek I, Nazar K (1988). Influence of fat-rich diet on physiological responses to prolonged physical exercise. In: Parizkova J (ed), *Charles University, Prague, Nutrition, Metabolism and Physical Exercise*, pp.179-193.

17. Grucza R, Pekkarinen H, Titov EK, Kanonoff A, Hanninen O (1993). Influence of menstrual cycle and oral contraceptives on thermoregulatory responses to exercise in young women. *Eur J Appl Physiol.* **67**:279-285.
18. Bar-Or O (1989). Temperature regulation during exercise in children and adolescents. In: Gisolfi CV, Lamb DR (eds.), *Benchmark Press, Carmel IN, Perspectives in Exercise Science and Sport Medicine: Youth Exercise and Sport*, pp. 335-362.
19. Kenney WL (1993). The older athlete: exercise in hot environment. *Sports Science Exchange* 44: 6, No3.

Self--evaluation questions for Chapter 8.1./2.

1) How large an increase in core temperature would you expect in a person exercising for 40 min at 50% of $\dot{V}O_{2max}$.

2) Calculate the maximal oxygen consumption of a person with maximal HR 188 and stroke volume 120 (maximal arterial oxygen content is 200 ml l^{-1}). The person is working with an oxygen uptake of 1.8 l · min^{-1}. What is the relative work load of the person (% $\dot{V}O_{2max}$)?

3) What is the heat production when the oxygen uptake is 2 l · min^{-1}? (what was the efficiency of the work in your calculation?).

4) Describe how much heat (in W) is transported to the skin, when the skin blood flow is 3 l · min^{-1} and Tre and Tsk are, respectively, 38.5°C and 35.5°C?

5) Why would a higher sweating rate and blood volume improve the ability for exercise performance in hot environments?

6) The prescriptive zone is more narrow in water than in air. Why?

7) How will dehydration by sweating affect themoregulation during exercise?

8) In what way does the menstrual cycle affect core temperature in women?

9) Is warming--up always beneficial for exercise performance?

Chapter 9.1

Learning Objectives
After reading this chapter, you should be able to achieve the following as they relate to the human neonate:

- Describe the thermoregulatory crisis than occurs for the neonate at birth

- List the factors which predispose the neonate to heat loss

- Describe the role of brown adipose tissue in the neonate

- Describe the mechanism of non-shivering thermogenesis as it occurs in the neonate

- Discuss how the neonate responds physiologically to cold and hot environments

- Designate what the limits of ambient temperature are within which the normal full-term neonate can maintain normal body temperature

- Suggest in which ways the pre-term neonate is at a thermal disadvantage

- Suggest appropriate incubator temperatures for normal full-term, and pre-term infants

- Argue that the normal neonate shows thermoregulatory competence within certain environmental temperature limits

- Describe what behaviours neonates use for thermoregulatory purposes

- List the prerequisites for fever development, and those which may be lacking in the normal neonate in the first days after birth

- State the WHO recommended minimum ambient temperature for neonates

- State two disadvantages of hypothermia in neonates

- State two possible advantages of hypothermia in neonates

- State the two major routes of heat loss from the fetus to mother

- Give a typical value for the difference in body temperature levels between fetus and mother

Bullets

B-1 The fall in body temperature that occurs at birth, were it to occur in an adult could be sufficient to be life-threatening.

B-2 The normal full-term neonate functions as a small, endothermic homeotherm.

B-3 The WHO recommends an ambient temperature for neonates of not less than 25°C.

B-4 Newborn babies cannot sweat at the same rate as adults, even when the rates are adjusted for body surface area.

B-5 Incidents of hyperthermia in babies are more often the result of overdressing and overcovering than of excessively high ambient temperature.

B-6 Diagnosis of fever in the newborn may be as difficult as it is important.

Chapter 9.1

THERMOREGULATION IN THE NEONATE

HELEN P LABURN
Department of Physiology and Brain Function Research Unit
University of the Witwatersrand, Johannesburg, South Africa

Introduction

The variability of body temperature in mammalian neonates is commonly thought to be due to immaturity of thermoregulatory processes. However, those processes we can measure and observe are evidence that neonates are, in fact, remarkably competent homeotherms. This chapter deals with questions such as how does the neonate cope with the thermal stress of birth, what thermoregulatory mechanisms are available to the neonate in the neonatal period, and how does the neonate acquire thermoregulatory competence before birth?

1. Birth: A Thermoregulatory Crisis

Immediately after birth, the body temperature of the human neonate falls rapidly, by between 1° and 3°C even in conditions controlled to provide a benign environment. Similar acute falls in body temperature occur in the neonates of other species, and are more severe for newborn delivered into cold conditions. Figure 1 shows the fall in body temperature that occurred in a human neonate and in a group of newborn lambs, born into similar room-temperature conditions. The rate of fall is greatest in the first few minutes after delivery and lasts up to several hours, even in babies given routine delivery-room care. Several factors contribute to the fall. When the fetus emerges, it has a body temperature of about 38°C, about 0.5° C above that of its mother, and consequently a very large gradient exists for heat loss to the surroundings. The body surface area-to-mass ratio of the neonate is two to three times that of an adult, so heat loss occurs with great facility down that gradient. Finally, evaporation of amniotic fluid occurs from the skin surface.

The fall in body temperature that occurs at birth, were it to occur in an adult, could be sufficient to be life-threatening. However, except when birth occurs into the harshest of cold environments, neonates apparently suffer no adverse sequelae to

B-1

the thermoregulatory crisis, and indeed are capable not only of halting the fall in temperature but reversing it. Although the recovery from hypothermia is more rapid in some animal species, the body temperature of the normal human neonate, even without active intervention, approaches that of an adult within 24 hours.

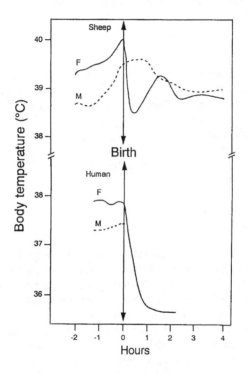

Fig. 1:Changes in body temperature occurring after birth in a group of lambs (top) and in a human neonate (bottom). Solid line in each graph indicates fetal (F)/neonatal body temperature. Interrupted line indicates maternal body temperature (M). Ordinate, body temperature in °C, and abscissa, time before and after birth, in hours, also indicated by vertical line with arrows corresponding to time zero. Reproduced with permission from the American Physiological Society (from Laburn, 1996) [1].

Recovery from the birth-related hypothermia requires a dramatic physiological response, and it comes in the form of a rise in metabolic heat production, mainly in brown adipose tissue (BAT, see below). Initiation of the life-saving metabolic response requires a combination of triggers, which all normally occur naturally at birth; cold-sensitive thermoreceptors in skin are stimulated as skin temperature falls below 35°C, blood oxygen tensions rise, and the neonate is separated from the placenta. Given sufficient BAT, which is laid down from the 26th week of human gestation, but only in adequate amounts from the 36th week, the postnatal hypothermia is counteracted by the stimulation of non-shivering thermogenesis (NST) in BAT, via the intact neonatal sympathetic nervous system. Babies who have any deficiency in the prerequisites for thermogenesis, such as inadequate oxygen supply, or insufficient BAT, have more difficulty reversing hypothermia in the first hours and days of life.

2. Adaptation to the Post-Birth Environment

2.1. Heat production

The normal full-term neonate functions as a small, endothermic homeotherm; that is, an organism capable of maintaining constancy of body core temperature primarily by means of endogenously produced heat. More thermogenesis per kilogram body mass is required in the neonate compared to the adult, because of the neonate's predisposition to heat loss; if body core temperature is to be maintained, that heat loss must be opposed. The greater heat loss from the newborn per unit surface area, arises primarily from passive heat transfer from the newborn's large surface area in relation to its mass, and poor insulation against that heat loss provided by skin and underlying tissues at birth. As a consequence, standard metabolic rate (SMR) in the neonate in the first days after birth (oxygen consumption is about 6 ml/kg.min) is approximately twice that per kilogram body mass of an adult in resting conditions. (Even in warm environments, the neonate is hypermetabolic, because of high levels of tissue growth and differentiation). The necessary thermogenesis occurs primarily in BAT which is found in several locations in the neonate, with major interscapular, periaortic and perirenal deposits.

B-2

Non-shivering thermogenesis (NST) in BAT results from the presence, in this tissue, of an uncoupling protein (UCP) that short-circuits protons in the process of oxidative phosphorylation, such that heat, rather than adenosine triphosphate (ATP) is produced (see also Chapter 9.2). The concentration of mRNA for UCP peaks before, at or just after birth, depending on animal species. Cold exposure stimulates beta-adrenergic receptors on BAT cells and the synthesis of UCP. BAT cells are rich in mitochondria with enzymatic machinery capable of catabolizing lipids, providing energy for the oxidative phosphorylation and the resultant heat. Brown adipose tissue also has a disproportionately large blood supply (about one fifth of cardiac output during thermogenesis), such that the heat generated is transported away from BAT to the body core. Body core warming is facilitated by the fact that BAT is strategically deposited so as to warm blood coursing through or to the body trunk. Heat production by BAT requires oxygen, which can be consumed at a rate of up to about 15 ml/kg.min in infants just a few days old exposed to room temperature conditions; this rate is similar to that (17 ml/kg.min) of healthy, fit young adults exposed to 5°C. In the adult human, the heat production results from shivering thermogenesis, because BAT disappears, as a thermoregulatory tissue at least, within the first year of life.

Can the neonate shiver as well as activating BAT? In newborn guinea pigs, shivering can be elicited when NST is blocked pharmacologically, but in the human, shivering is seen only in the severely hypothermic infant. Thus, shivering may be either suppressed by NST, or brought into play only once body core temperature has fallen to levels much lower than that necessary to trigger NST.

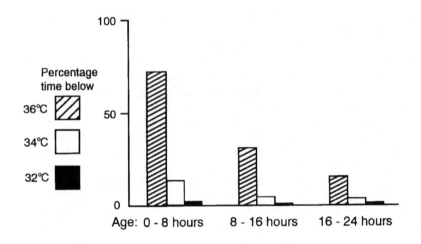

Fig. 2: Percentage of time spent with body temperatures below 36°C, 34°C or 32°C (ordinate) in a group of 35 normal human neonates born into normal delivery room conditions and subsequently moved to a nursery ward, within the first 24 hours after birth. Abscissa shows the ages of the neonates having body temperatures below the three cut-off temperature levels. Neonates 0-8 hours old had body temperatures below 36°C about three-quarters of the time, body temperatures below 34°C about 15% of the time, and some of these neonates had body temperatures below 32°C for 2% of the time. Neonates 8-16 hours and 16-24 hours old spent a lower percentage of their first hours having body temperatures below each of the cut-off temperatures. (Data from Ellis *et al*, 1996) [2].

Though a powerful thermogenic process, BAT metabolism may be inadequate to reverse the hypothermia that results when neonates are exposed to environmental temperatures below 23°C. Figure 2 shows the results of 24-hour ambulatory temperature monitoring of human neonates from 90 minutes after birth into ward conditions where ambient temperature was 22.5°C on average. The babies were all dried and wrapped, but for 85% of the time were what the World Health Organisation (WHO) classifies as "moderately hypothermic" (32° - 36°C). Some babies were severely hypothermic. The WHO recommends an ambient temperature for neonates of not less than 25°C.

B-3

2.2. Heat Loss

Most neonates are born into conditions in which heat loss occurs by all four available routes: conduction, convection, radiation and evaporation. Transdermal (insensible) water loss is considerable in neonates, especially in preterm babies, in whom the skin barrier consists of a few cell layers only. Non-evaporative heat loss decreases as ambient temperature rises, and when that temperature exceeds about 35°C, the neonate, even of only a few hours' age, will invoke active evaporative heat loss by means of sweating (Fig. 3A). Both an increase in body core temperature to above 37.5°C and an increased skin temperature to about 36°C, are required before sweating commences, and then it does so in a quite sudden fashion. The temperature thresholds for sweating are higher than those in adults, in whom sweating may occur once average skin temperature has reached only 34°C, and ambient temperature only 30°C. The body core temperature threshold for sweating onset falls within the first days of life (Fig. 3B), and is highest in preterm infants, except that babies born more than six weeks prematurely do not sweat at all.

Newborn babies cannot sweat at the same rate as adults, even when the rates are adjusted for body surface area, and in neonates, the distribution of functional sweat glands differs from that in the adult. At maximum sweating capacity, about 45% of sweat in neonates is produced by glands in the skin of the head, particularly the forehead, even though the head constitutes only one fifth of the body's surface area. Sweating is the only active means of increasing heat loss when ambient temperature exceeds body temperature, and in infants, just as in adults, obstructing evaporation can have dire consequences. Covering the head of a baby, for whatever reason, will obstruct the most effective way of actively losing heat in warm circumstances, as will placing the baby in a prone, rather than in a supine position on bedding. Hyperthermia has been implicated as a cause of sudden infant death syndrome.

B-4

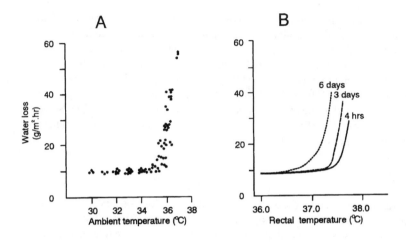

Fig. 3: Rates of evaporative water loss, in g/m².hr (ordinate) measured in **A**: a group of three human neonates exposed to rising ambient temperature (abscissa) on occasions during the first 9 days after birth. Some of the measurements were made on the first day of life. In **B**: relationship of water loss to increase in rectal temperature (abscissa) as measured in a neonate, when 4 hr old, 3 days old and 6 days old. Notice that the threshold for onset of sweating declines with age, even in the first few days after birth. Redrawn from Hey & Katz, 1969 [3].

2.3. Vasomotor responses

In both full-term and premature infants, well-developed vasomotor responses to changing ambient and body temperature have been observed, from the second day after birth, and in some infants on the first day of life. As for sweating responses, the body temperature threshold for vasodilatation is higher than it is in adults. Similarly, body temperature must fall lower in the newborn to elicit the opposite response, vasoconstriction. Local cooling or warming, especially of the face, can elicit vasomotor responses at local or distant body sites. However, even more so than in adults, the full range of vasomotor responses occurs over a narrow range of body temperature. In the human neonate, that range of body temperature is 36.5 - 37.5°C and may be even narrower in some infants.

2.4. Behavioural responses

Behavioural responses to thermal stimuli often are treated as Cinderella responses in human thermoregulation. Behaviours in hot or cold conditions are stimulated probably by the same sensory inputs as are autonomic responses, and the thresholds for initiating appropriate responses indeed may be lower than for some autonomic effector mechanisms. Except in situations where personal modesty does not allow it, most adults initiate behavioural activities preferentially, rather than allowing body temperature to rise or fall such that shivering or sweating must be invoked (see also Chapter 7).

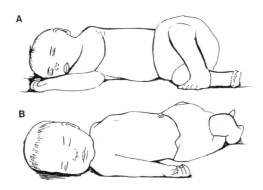

Fig. 4: Diagramatic representation of the postures of a 3-day old infant in an incubator, the temperature of which was 31°C and rectal temperature less than 37°C (**A**) and **B**, when incubator temperature was 37°C and rectal temperature 37.7°C. At higher environmental and body temperatures the infant adopted an extended, "sunbathing" posture, which allows for greater surface heat loss. Drawn from Harpin et al,1983 [4].

There is every reason to believe that human neonates too possess behavioural mechanisms for thermoregulatory control of body temperature. Certainly, the precocial newborn of several animal species clearly use behavioural thermoregulation in natural or experimental conditions. The behaviour takes the form of postural changes, huddling, or seeking and moving to a warmer or colder environment. Newborn infants, irrespective of gestational age, become less active at high ambient temperatures, and adopt an extended rather than a flexed body position (Fig. 4). The response occurs readily even in premature infants, but is most consistent in full-term infants, and after one week of extra-uterine life. Other forms of behaviour the human newborn uses include restlessness and crying at both very high- and low ambient temperatures. Crying increases heat production, and initiates responses in parents too. It seems reasonable to assume that the thresholds for triggering thermoregulatory behaviour in neonates would be shifted outwards, compared to those of adults, as are the autonomic thresholds.

2.5. Implications for the care of neonates

The limits within which the human newborn can regulate its body temperature unassisted are defined largely by its heat-generating and conserving, and heat-losing capabilities, and are considerably narrower than those of the adult. The maximum (summit) metabolic response in a neonate occurs at an ambient temperature of about 23°C; in a similarly clad adult, at about 5°C. The neonate can sustain summit metabolism only for about an hour; thereafter, body temperature cannot be maintained. Upper limits of ambient temperature control for the neonate are difficult to predict; incidents of hyperthermia in babies are more often the result of overdressing and overcovering the baby than of excessively high ambient temperature. At high temperatures, the neonate can dissipate heat by sweating only to the equivalent of its standard metabolic rate; adults can dissipate about five times their standard metabolic rate through sweating. Except when relative humidity is high or the neonate inappropriately clothed, the human neonate probably can withstand ambient temperatures up to and equal to its body core temperature, largely because of its inherent predisposition to lose heat passively.

B-5

Interventions for maintaining neonatal body temperature constant should be aimed at keeping the body temperature between 36.0° and 37.5°C. Especially in the preterm infant, who is much more thermally labile than a full-term infant, but also for small full-term neonates, incubator nursing is desirable if not essential, unless the nursery temperatures are sufficiently high. Avoiding ambient temperatures that are too low for the premature infant may not be possible in some communities (see for example the study referred to in Fig. 2), but the consequences are increased numbers of perinatal deaths. Incubator temperatures should be regulated so as to maintain the infant in its thermoneutral zone, that ambient temperature range where both heat production (through thermogenesis) and heat loss (via sweating) are at a minimum (see Chapter 2). For the full term infant, this thermoneutral zone is a narrow zone around 34°C, and it should be raised 0.5°C for every two weeks of prematurity, as a rough guide. As the baby increases in mass, even in the first days after birth, but particularly beyond that, skin tissue insulation increases, the body surface area-to-mass ratio decreases, and so does heat loss. Thus the thermoneutral zone falls and widens, eventually to between 26 and 28°C in naked adults.

Does hypothermia matter? Clearly, severe hypothermia can itself be fatal in neonates, as in adults. Moderate hypothermia of a chronic nature in the post-birth period is associated with increased mortality and morbidity in the newborn, while milder hypothermia makes greater metabolic demands on the neonate, and leads to decreased growth. The hypothermia inevitable at birth may, on the other hand, have some advantages: Thus, it may enhance the initiation of normal lung function in the newborn; also, the higher the brain temperature the greater the risk of severe neuronal damage in the peri-partum period, if the baby were to become hypoxic, for example during prolonged labour.

3. Fever in the Newborn

Fever is one of the first signs of infection or inflammation. However, in the first few days after birth, the infected neonates of many species, including humans, do not always develop adult-like fevers. Thus early diagnosis of an infection in the neonate may be as difficult as it is important.

The development of fever requires several stages: the immune cells of the body must recognise and react to the infectious agent, and produce mediator substances (eg cytokines). Thereafter, the mediator substances, directly or via other agents (probably prostaglandin), in turn act on the hypothalamus to induce appropriate heat-conserving responses and thermogenesis to produce the rise in body temperature (see Chapter 10). The anomalous febrile response in neonates is not well understood, but seems to have to do with the inability to recognize certain pyrogens, especially Gram-negative bacteria or their products, and to mount the appropriate release of cytokine mediators in response to the bacterial challenge, for up to four days after birth. Cytokine production by monocytes from human neonates may differ from that in adults, such that not all the prerequisite, fever-inducing cytokines are produced in response to Gram-negative bacterial stimulation of neonatal monocytes, and moreover, significantly higher concentrations of a cytokine receptor antagonist (interleukin-1 ra) are present in the neonate's circulation. There may be endogenous antipyretic agents, such as arginine vasopressin and alpha-melanocyte stimulating hormone, present in high concentrations in neonates too, which also could play a role in the suppression of fever. Finally, there may be a relative insensitivity of the hypothalamic centres to the various mediators involved.

4. The Ontogeny of Thermoregulation

The neonate, which emerges into a dry, cool, air environment, developed as a fetus in very different circumstances - a fluid medium with a temperature in the region of 38°C. How, then, is it possible that, at birth, a neonate is so extraordinarily well-equipped physiologically to deal with cold (and hot) ambient conditions to which it has had no prior exposure? Current evidence, based largely on work in fetal lambs, implies that thermoregulatory sensory and effector pathways develop under genetic cues, as do most other physiological functions *in utero*, and not because of adaptive responses to temperature fluctuations that may occur around the fetus, nor because the fetus has opportunities to practice thermoregulation in response to changes in its environmental conditions.

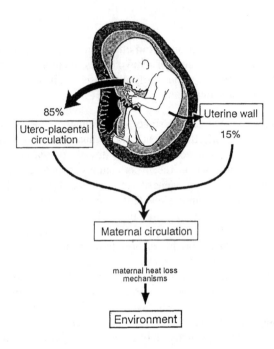

Fig. 5: The fetus produces considerable metabolic heat, which must be dissipated via maternal structures. Most fetal heat is lost via exchange with cooler maternal blood at the placenta, and a smaller proportion of heat is lost via exchange across the amniotic space and uterine wall. Ultimately the fetal heat is exchanged with the environment through normal maternal heat loss mechanisms. Reproduced with permission from the American Physiological Society (from Laburn, 1996) [1].

The body temperature of fetal lambs at least, and probably of the fetuses of other animals of similar body mass, is remarkably constant, varying with an amplitude of only 0.3°C during any 24-hour period in a 20 -25°C environment. During this period, much larger circadian excursions of body temperature occur in the mother animal. In the last six weeks of gestation, despite large gains in body mass, body temperature of the fetal lamb changes less than 1°C. Moreover, when the mother is exposed to heat and cold stresses, including exercise, the fetus' inherent thermal inertia and thermoregulatory strategies the mother uses for her own thermoregulation, collaborate to reduce changes in the fetal body temperature. Some of these strategies may involve varying blood flow to the placenta, the main site of heat exchange between fetus and mother (Fig. 5); during heat stress, blood flow to and from the placenta may be increased, as it is to many other peripheral tissues, so that more heat is lost from the fetus. Failure to dissipate heat adequately leads to intra-uterine hyperthermia, a condition associated with a variety of congenital abnormalities.

Because the fetus has no access to an air environment of its own for dissipation of body heat, there is no apparent advantage in it displaying thermoregulatory autonomy yet, it requires thermoregulatory competence immediately after birth. It would appear that at least some of the thermoregulatory mechanisms needed in the post-birth environment are indeed competent *in utero*, but are suppressed . After a surgical procedure similar to a Caesarian section, allowing for simulation of the neonatal state after the fetal lamb is returned to the uterus, cooling of the late-gestation fetus results in the onset of non-shivering thermogenesis. Indeed, it would appear that a substance or substances from the intact placenta acts as a thermoregulatory inhibitor during gestation. At birth, when the cord is cut, the neonate's thermoregulatory mechanisms are freed from this suppression and come into play as and when the thermal stimuli so require.

Acknowledgements

I am grateful to Duncan Mitchell for helpful comments on the manuscript.

References

1. Laburn HP (1996). How does the fetus cope with thermal challenges? *News Physiol Sci* **11**: 96-100.

2. Ellis M, Manandhar N, Shakya U, Manandhar DS, Fawdry A and de L Costello AM (1996). Postnatal hypothermia and cold stress among newborn infants in Nepal monitored by continuous ambulatory recording. *Archiv Dis Child* **75**: F42-F45.

3. Hey EN and Katz G (1969). Evaporative water loss in the new-born baby. *J Physiol* **200**: 605-619.

4. Harpin VA, Chellappah G and Rutter N (1983). Responses of the newborn infant to overheating. *Biol Neonate* **44**: 65-75.

Suggested Reading

Bolton DPG, Nelson EAS, Taylor BJ and Weatherall IL (1996). Thermal balance in infants. *J Appl Physiol* **80(6)**: 2234-2242.

Brück K (1961). Temperature regulation in the newborn infant. *Biol Neonat* **3**:65-119.

Brück K (1992). Neonatal thermal regulation. In: Polin RA, Fox WW (eds), *Fetal and Neonatal Physiology*. WB Saunders Company, Philadelphia, pp. 488-515.

Laburn HP, Mitchell D and Goelst K (1994). From fetus to neonate - implications for the ontogeny of thermoregulation. In: Milton AS (ed), *Temperature Regulation Advances in Pharmacological Sciences*. Birkhäuser Verlag, Basel, pp.229-240.

Power GG (1992). Fetal Thermoregulation: Animal and Human. In: Polin RA, Fox WW (eds), *Fetal and Neonatal Physiology*, WB Saunders Company, Philadelphia, pp. 477-483.

Sauer PJJ (1991). Neonatal Thermoregulation. In: Cowett RM (ed), *Principles of Perinatal-Neonatal Metabolism*, Springer-Verlag, New York, pp. 609-622.

Stanier MW, Mount LE and Bligh J (1984). Temperature regulation in the newborn and in old age. In: Comline RS, Cuthbert AW, Dixon KC, Herbert J, Iversen SD, Keynes RD and Kornberg HL (eds), *Cambridge Texts in the Physiological Sciences 4. Energy balance and temperature regulation*. Cambridge University Press, Cambridge, pp.73-85.

Webster MED (1987). Temperature Regulation in Children. In: Shiraki K, Yousef MK (eds), *Man in Stressful Environments Thermal and Work Physiology*. Thomas, Springfield, pp. 35-43.

Review questions

1. What changes occur in body core temperature of the neonate in the immediate post-partum period? What factors contribute in a substantial way to these changes?

2. How long does it take typically for the human neonate's body temperature to reach levels similar to that of an adult, in the immediate post-partum period?

3. What is the peculiar property of BAT?

4. During what period of gestation in the human, is BAT deposited?

5. Describe the phenomenon of non-shivering thermogenesis.

6. By how much more is human neonatal standard metabolic rate greater than that of an adult's?

7. At what ambient temperature does a lightly clad full-term neonate generate summit metabolism? At what temperature does a similarly clad young adult generate summit metabolism?

8. How does the WHO classify "moderate hypothermia" in neonates? What is the WHO's range for "severe hypothermia"?

9. At what age after birth does the sweating mechanism become operative in the human?

10. What occurs to the body core temperature threshold for onset of sweating as the newborn baby ages?

11. Describe what changes occur in evaporative water loss in newborn babies exposed to increasing environmental temperatures between 30° and 38°C.

12. What parts of the body contribute most to sweating in the human neonate?

13. What is the body core temperature range in a full-term neonate within which changes in skin temperature will elicit vasomotor changes?

14. What behaviours are observed to occur in the newborn of certain animal species in response to temperature changes in the environment?

15. What thermoregulatory behaviours are observable in the human neonate?

16. What is meant by "the thermoneutral zone"?

17. What would be an appropriate incubator temperature for an infant born at 36 weeks of gestation? Why would a lower incubator temperature be required to maintain a constant body temperature in a baby born at full-term?

18. Why may infection be difficult to diagnose in the neonate?

19. What aspects of the adult fever response may be lacking in the neonate?

20. The fetus is deprived of an air environment into which to dissipate its metabolic heat. How does it manage to maintain thermal homeostasis?

Chapter 9.2

Learning Objectives
1. Understand that "physiological" and "chronological" age may not always coincide.
2. Understand that diseases which cause unconsciousness, or immobility, predispose to thermoregulatory disorders.
3. Understand that damage to thermoregulatory centres and pathways, and to peripheral autonomic nerves, predisposes to heat and cold illness.
4. Understand that diseases which impair thermoregulation are commoner in old age.
5. Be aware that multiplicity of medication may put an old person at risk of impaired thermoregulation, and that some drug actions may alter with advanced years.
6. Understand that a significant number of old people may have impaired perception of temperature, impaired shivering, impaired peripheral vascular control or a combination of these.
7. Know that socio-economic factors may play a role in thermoregulatory disorders in the elderly.
8. Utilize basic knowledge of thermoregulatory physiology and physics to predict the risk of factors and principles of management for thermoregulatory disorders in the elderly.

Bullets
B-1   Physiological and chronological ages.
B-2   Brain micro-infarcts.
B-3   Peripheral circulation.
B-4   Metabolic rate in the elderly.
B-5   Evaporative cooling.
B-6   Temperature sensation.
B-7   Effects of age on fever.
B-8   Neurological disorders and thermoregulation.
B-9   Effects of drugs on thermoregulation.
B-10 Socio-economic factors in thermoregulatory disorders.
B-11 Acclimatization to heat in the elderly.

Chapter 9.2

THERMOREGULATION IN THE ELDERLY

KEITH E. COOPER
Department of Medical Physiology, Faculty of Medicine
University of Calgary, 3330 Hospital Drive N. W.
Calgary, Alberta, Canada. T2N-4N1

1. Introduction

All too often the term "elderly" is used as though there were homogeneity of function in all persons above an arbitrarily defined age. There are two views of ageing, the first adhering solely to the chronological age and the second to evidence of physiological changes related to advancing years. The two do not always coincide. There are people of advanced years chronologically who are functionally as well preserved as many in much younger age groups; and there are some relatively young persons whose overall body functions belong to an average group of much riper years. So if we refer to the elderly as those of 65 years and older, by arbitrary decision, we must avoid the temptation to assume a physiologically homogeneous population. While there is a trend for disease processes to occur with increasing frequency, or for deteriorating body function to occur in the later years, there remain many older persons with remarkably well preserved bodily function.

B-1

We must also attempt to disentangle the possible alterations in thermoregulatory functions to factors purely related to physiological ageing from those related to diseases which occur more frequently as the years progress. As will be discussed later, one factor in the development of heat illness may be obesity, and this tends to occur in people who have grown older, become more affluent and have taken inadequate exercise. Immobilisation due to arthritis or strokes reduces the capacity for heat production by exercise, or by moving about to adjust room thermostats, etc, and thus may predispose to hypothermia. Reduced perception of cold is not uncommon in the elderly, and this may also predispose to reduced body temperature.

Autopsy examinations of brains reveal small infarcts in many areas, and these infarcts can be found in brains of quite young people. They are usually too small to cause any observable changes in neurological function. Should such micro-infarcts occur in the small hypothalamic region, which contains thermoregulatory neurons as

B-2

a small proportion of the total neuron population, then thermoregulatory deficits might occur. In much older people the likelihood of such infarcts occurring is higher but it is still not possible definitely to correlate their discovery at autopsy with thermoregulatory problems.

With these preliminary comments, we can now examine more closely some of the thermoregulatory deficits reported to occur in older people.

2. Thermoregulatory Changes Reported to Have Occurred in the Elderly

2.1. Peripheral blood flow

One study of peripheral blood flow responses to body warming in elderly patients who had no overt signs of disease, and whose ages ranged from 68 to 94 years, indicated that raising the core temperature or applying radiant heat to the trunk evoked hand vasodilatation. The vasodilatation was, with the exception of a 94-year old man, of the same order as that reported for young adults. In this man, the responses were present but reduced [1]. However, in a more detailed study of 64 healthy elderly people [2], it was found that there was significant diminution of some autonomic nervous functions including reduced vasoconstriction in response to body cooling. The reflex vasodilatation in the hand evoked by radiant heat applied to the trunk, which normally is inhibited at body temperatures below 36.6°C, still occurred at much lower body temperatures in elderly survivors of hypothermia [3]. In a hot environment (38°C) with moderate humidity, men in the 41-57-year age range tended to have higher forearm blood flows at rest, and the increase in blood flow during work was greater in the older age group [4]. The higher blood flows in the forearms only occurred in very hot environments and not when exercising at 26°C environmental temperature. The increased skin blood flow, occurring under conditions in which heat loss was achieved entirely by evaporation, did not contribute to greater thermoregulatory efficiency and became an unnecessary strain on the cardiovascular performance of the older men in the heat. Although the "older" group could not be considered as being of advanced years, the responses observed do indicate an age-related alteration in one type of heat-induced peripheral blood flow response. Some alterations in the control of the skin blood flow in response to changes in environmental temperatures have thus been demonstrated in people of greater ages, but the altered responses are not universal, and the underlying physiological or pathophysiological causes of these changes remain obscure.

2.2. Shivering and heat production

It is known that the metabolic rate falls with the advance of old age [5], and this is in part due to a smaller proportion of heat-producing cells in the total body mass. As people get older they tend to be more sedentary, and this also reduces the metabolic rate. A significant number of elderly people fail to shiver during cold exposure [6]. Others, in their studies of elderly survivors of hypothermia, [3,7] found reduced or absent shivering in response to body cooling. However, the universality of reduced shivering in response to cold exposure among an aged population is far from established. Whether brown adipose tissue contributes to body heat production in the human adult is still controversial, but it is likely that the numbers of dispersed brown fat cells, as may persist in adults, would be greatly reduced in old age (see also Chap. 4.2.). B-4

2.3. Evaporative heat loss

There is good evidence that the sweating responses to both raised body temperature and local chemical stimulation of the sweat glands are reduced in those over 65 years old as compared to those of lesser age. This is of great importance during severe heat waves, in which evaporative heat loss is the main route of cooling, if socio-economic factors prevent the use of air-conditioners. It has also been shown to be a major factor in the development of heat illness in older pilgrims attending the great Muslim Makkah Hadj at which, depending on the dates of the pilgrimage, the dry bulb temperatures may vary from 35-50°C and the radiation heat load is high [8]. Many pilgrims come from places where diseases which affect work capacity and thermoregulation are common, and the physiological age may be greater than the chronological age. B-5

2.4. Perception of environmental temperature

As Collins and Exton-Smith [5] pointed out, many types of sensory perception show some deterioration in advanced age. Hearing loss is, apart from the cochlear damage caused by the extreme noise of rock concerts and portable CD players, much more common in the elderly, and some skin sensory functions are also blunted. Among these modalities, skin thermoreceptors seem less able to discriminate temperature intervals, or perhaps the central processing of normal peripheral thermal information is altered in old age. The ability of old people to manipulate their environmental B-6 conditions is, on the average, less precise than it is in the young adult. This could be one reason for the increased susceptibility to hypothermia in the elderly.

3. Evidence from Animal Experiments

There is evidence that the brain concentrations of two neurotransmitters which are used in thermoregulatory pathways are reduced in old animals. These transmitters are acetylcholine and norepinephrine. Further, old rabbits seem to be more sensitive to the topical application of these substances in thermoregulatory regions than are young animals. Fever can be produced in the rabbit by intravenous administration of endotoxin. Old rabbits have smaller fevers produced in this way as compared to young animals, and this diminution of fever is greatest in the second peak of a biphasic fever. During the fever caused by endotoxin in the older rabbits, there is a large rise in the plasma epinephrine level. Some recent evidence suggests that, since epinephrine suppresses the production of some cytokines, the rise in circulating epinephrine in the older animals could be a factor in reduced fever in the older animals. It is also possible that the change in fever responses in the old animals may be related to as yet unidentified changes in the neuronal structure of the thermoregulatory pathways. While the evidence for this remains somewhat vague, there is enough to warrant further detailed research into the afferent and efferent regulatory systems in young and old animals. There could also be some down-regulation of pyrogenic cytokine receptors or thermoregulatory transmitter receptors, but again this is still mainly in the realm of speculation.

Diazepam, a tranquilizing drug often used by elderly people, has been found to cause hypothermia to a slightly greater degree in old as compared to young monkeys, [9]. This was especially so in cold environments. Also, diazepam given into a lateral cerebral ventricle caused a smaller hyperthermia in old as compared to young animals. In the case of the hypothermia, the drug effect appeared to be mainly peripheral. There is thus an inherent danger of hypothermia in the use of this drug in elderly humans exposed to cold. Many other drugs present similar risks (see later).

Old mice have been found to become hypothermic on cold exposure, and hyperthermic during heat exposure more easily than young mice; the heat response is not humidity-related [10].

4. Clinical Aspects of Thermal Malfunctions in the Elderly

4.1. Neurological disorders

4.1.a. Strokes
As a result of a stroke, the patient may be immobilized and thus have a lowered metabolic rate. If multiple infarcts have occurred and one or more damage the hypothalamic thermoregulatory regions or the thermoregulatory efferent pathways in the brainstem, then temperature regulation will become inaccurate or, in extreme cases, absent.

B-8

4.1.b. Organic brain diseases such as Alzheimer's disease
In such cases the problem lies in the loss of the ability to reason and adjust clothing or thermostats, even if the pathways for autonomic thermoregulation remain intact. The patient may forget the task which he/she set out to do in order to manipulate the environmental temperature.

4.1.c. Hypothalamic disorders
The thermoregulatory mechanisms in the hypothalamus can be damaged or destroyed by tumors or diseases involving diffuse cellular infiltration. Such disorders may occur throughout life and are not more common in old age.

Vascular disease leading to infarction of small areas in the hypothalamus are, however, more likely as age increases. These hypothalamic disorders of thermoregulation are often accompanied by signs of involvement of other body systems due to damage to extrahypothalamic regions. Even with pure hypothalamic lesions, there may be a combination of thermoregulatory disorders with endocrine disturbances, and sometimes the thermoregulatory and endocrine functions may be altered independently. One of the endocrine disturbances which predisposes to hypothermia without necessarily involving damage to neural hypothalamic mechanisms is hypothyroidism. In this case, the problem lies principally in the low metabolic rate. Hypopituitarism may also predispose to hypothermia even though the patient is not obviously myxedematous. This author has seen a patient with unrecognized hypothyroidism who developed a myxedema psychosis for which he was treated in a psychiatric institution with chlorpromazine. The resulting hypothermia was profound!

4.1.d. Impairment of autonomic nerve function
These is evidence for some deterioration of peripheral sympathetic nerve function in advanced age. Sometimes this is associated with long standing diabetes mellitus. In others, it may be associated with prolonged alcohol abuse. There may be thiamine deficiency as a result of alcoholism, which could lead to small petechial hemorrhages in the hypothalamus and elsewhere in the brain (Wernicke's encephalopathy). Since the control of skin blood flow and sweating depend on efferent sympathetic nerves, peripheral autonomic neuropathy from any cause can blunt these important thermoregulatory functions.

4.1.e. Coma
This may be the result of diabetes mellitus, cerebral vascular accident, alcohol, or trauma, to mention but a few causes. The loss of consciousness deprives the patient of the ability to use behavioral thermoregulation. Depending on the cause of the coma, the physiological thermoregulatory control systems may also become disrupted. Again, none of these causes is confined to the elderly, but all may occur in old people.

4.1.f. Degenerative lesions which can affect the thermoregulatory outflow pathways in the spinal cord and brain stem
These include disorders such as intermediolateral column degeneration which can occur in both young and old persons, and Shy-Drager syndrome, a multiple system degeneration which occurs more frequently in older patients. Susceptibility to hypothermia is increased in Parkinson's disease. This disorder is more common in elderly than in younger people.

4.2. Immobility

Immobilization by such disorders as severe arthritis or paralysing diseases may, by reducing heat production, in combination with reduced thermal perception in the elderly, predispose to hypothermia. Some forms of arthritis occur often in young people, e.g., rheumatoid arthritis, while others, e.g., osteoarthritis are more common in the elderly. Again the lack of mobility may interfere with tthe physical effort necessary to adjust air-conditioning, etc., and in those living alone can lead to both cold and heat illnesses.

4.3. Drugs

The elderly, often smitten with multiple health problems, all too often take far too many prescription (and sometimes over-the-counter) drugs, particularly when in the care of several specialists between whom there is little communication. Some drugs which have been used to treat high blood pressure may, for example, by altering the performance of efferent sympathetic nerves, or altering hypothalamic function, blunt the central control of thermoregulation and the peripheral vascular and sweat responses in the elderly. Many more studies on the interaction of such drugs and thermoregulation are needed before the extent of this potential danger can be defined. Similarly, the effects of mood altering drugs on thermoregulation in the human would benefit from a great deal more study; there is evidence that some may depress thermoregulation in the elderly. Excess alcohol intake may lead to hypothermia and heat illness in the elderly, as well as in younger persons.

B-9

4.4 Malnutrition

This may occur as a result of poor socio-economic circumstances or as the result of confusion and mental impairment in some old people. Loss of insulation and muscle bulk are principally responsible for the impaired resistance to cold exposure. Another form of malnutrition, obesity, often caused by overeating an unbalanced diet and a sedentary lifestyle, can lead to a high risk of heat illness during severe heat waves.

B-10

4.5. Other consequences of poverty

One interesting cause of hypothermia in an economically challenged group was reported as due to lead poisoning. This was due to the cheap purchase or scrounging of discarded automobile batteries for burning as fuel in indoor wood-burning stoves. Inability to afford the means of heating rooms in cold weather, and to cool them during heat waves, are known causes of many temperature-related deaths in a number of the more affluent countries.

4.6. Simple acute exposure to cold

This may occur when a confused elderly individual attempts to walk to his/her destination in very cold weather. It may also occur by sitting in a poorly heated house, especially if the perception of cold is blunted.

4.7. Heat edema

This occurs commonly in the elderly and is manifested by peripheral pitting edema after a few days of exposure to high environmental temperatures. The cause is not clear, but the condition has to be differentiated from edema resulting from heart disease [11].

4.8. Syncope

Heat syncope occurs in susceptible persons who stand or work in hot environments. In the elderly, it is often associated with the taking of adrenergic β-receptor-blockers rendered necessary by cardiovascular disease. It may also be associated with peripheral autonomic neuropathy.

5. The Problem of Acclimatization in the Elderly

The literature on the effects of ageing on acclimatization to heat in the elderly is sparse and inconclusive. Early longtitudinal studies suggested that fit elderly individuals were able to acclimatize to heat, and to work well in hot environments. A more recent cross-sectional study showed that older men, though able to acclimatize to some extent to heat, did not have the same level of physiological improvement as younger subjects [14]. Table 1 presents some non-quantitative material on thermoregulation in the heat at various ages. While precise statistical data are not available, the information in the table is meant to draw attention to possible, theoretical changes in performance in the heat, and to suggest some causes of reduced performance.

THERMOREGULATION IN THE HEAT

Age (years)	Good	Less efficient	Poor
25	Most individuals. Greatly enhanced by acclimatization to heat.	Some individuals with obesity, lack of fitness, disease, unacclimatized to heat.	Few individuals. Unacclimatized, severe cardiac or respiratory disease; high spinal cord injury, congenital absence of sweat glands, etc.
50	Very many, especially heat-acclimatized individuals.	An increased number of unfit, obese, diseased, and unacclimatized individuals.	More unfit, unacclimatized individuals, some with compromised cardiovascular, respiratory, or CNS systems.
75	Some fit individuals with "normal" thermoregulation and capacity for a measure of acclimatization to heat.	More individuals with reduced cardiovascular and respiratory capacity for exercise, and reduced capacity for heat acclimatization. Some diseases, such as peripheral autonomic neuropathy.	Increased incidence of disease - e,g., cardiovascular, respiratory, or neurological, affecting capacity to thermoregulate or to acclimatize to heat.
80 +	Reliable data are not available at present (1997)		

Table 1. Some theoretical effects of ageing on thermoregulation in the heat.

6. Summary

It seems clear that some of the body temperature regulating mechanisms are altered with advanced years. There may be deterioration of the function of peripheral sympathetic nerves, and there is evidence for altered function of tbe brain thermoregulatory mechanisms. Some of these changes appear to be the consequence of old age per se. There is also alteration of the ability to perceive temperature change of the environment in many old people. In addition, there are disease processes which compound the pure effects of increasing age, such as atheromatous obstruction of arteries and micro-infarcts within the brain, which are more common in the elderly, in addition to major system disorders, such as strokes and Parkinson's disease. The frequent ingestion of multiple drugs by old people can also put them at risk of thermoregulatory disorders. It is important when attending the aged to bear in mind the possibility of such disorders and to test for them.

Some of the effects of cold exposure can be the same in the elderly as in younger people, for example cold allergies, diminution of manual dexterity, etc. One recent paper showed seasonal variation in blood pressure as heightened in old people, with blood pressure inversely related to environmental temperature [12]. This may be of importance in considering the higher incidence of cardiovascular accidents among the elderly in the winter.

Finally, it is also important to bear in mind that a substantial number of the elderly have normal thermoregulatory responses to challenges from alterations in the thermal environment.

References and Readings:

[1] Cooper, K. E. (1970). Studies of the human central warm receptor. In: Physiological and Behavioral Temperature Regulation. Eds. Hardy, J.D., Gagge, A. Pharo, & Stolwijk, J. A. J. Charles C. Thomas. Springfield, Ill. U.S.A.
[2] Collins, K. J., Exton-Smith, A. N., James, M. H., & Oliver, D. J. (1980). Functional changes in autonomic nervous responses with ageing. Age & Ageing. 9. 17-24.
[3] MacMillan, A. L., Corbett, J. L., Johnson, R, H., Smith, A, C., Spalding, J. M. K., & Wollner, L. (1967). Temperature regulation in survivors of accidental hypothermia of the elderly. Lancet 2, 165-169.
[4] Hellon, R. F., & Lind, A. R. (1958). The influence of age on peripheral vasodilatation in a hot environment. J. Physiol. (Lond). 141. 262-272.
[5] Collins, K. J., & Exton-Smith, A. N. (1983). Thermal homeostasis in old age. J. Amer. Geriatr. Soc. 31. 519-524.
[6] Collins, K. J., Dore, C., Exton-Smith, A. N., Fox, R. H., McDonald, I. C., & Woodward, P. A. (1977). Accidental hypothermia and impaired thermoregulation in the elderly. Brit. Med. J. 1. 353-356.
[7] Johnson, R. H., & Park, D. M. (1973). Intermittent hypothermia. J. Neurol. Neurosurg. Psychiat. 36. 411-416.
[8] Khogali, M., & Hales, J. R. S. Eds. (1983). Heat Stroke and Temperature Regulation. Academic Press. Sydney, New York.
[9] Clarke, S. M., & Lipton, J. M. (1981). Effects of diazepam on body temperature of the aged squirrel monkey. Brain Res. Bull. 7. 5-9.
[10] Hoffman-Goetz, L., & Keir, R. (1984) Body temperature of aged mice to ambient temperature and humidity stress. J. Gerontol. 39. 547-551.
[11] Harchelroad, F. (1993). Acute thermoregulatory disorders. Geriatr. Emerg. Care. 9. 621-639.

[12] Woodhouse, P. R., Khaw, R. T., & Plummer, M. (1993). Seasonal variation of blood pressure and its relationship to ambient temperature in the elderly population. J. Hypertension. 11(11): 1267-1274.
[13] Hardy, J. D., Gagge, A.P., & Stolwijk, J. A. J.   Charles C. Thomas Springfield, Ill. USA. pp 224-230.
[14] Yousef, M. K., (1987), In "Heat stress - Physical exertion and environment." Eds. Hales, J.R.S., & Richards, D.A.B. Excerpta Medica, Amsterdam, New York, Oxford. pp.367-382.

Additional Reading:
Horvath, S. M., Radcliffe, C. E., Hutt, B. K., & Spurr. G. B. (1955). Metabolic responses of old people to a cold environment. J. Appl. Physiol. 8. 145-148.

Paton, Bruce C., (1983). Accidental Hypothermia. International Encyclopedia of Pharmacology and Therapeutics: Thermoregulation; Pathology, Pharmacology, and Therapy. Eds. Schönbaum E., & Lomax P. Pergammon Press. New York, Oxford.

Self-evaluation questions for chapter 9.2

1. List 5 categories of disorders which could predispose old persons to hypothermia.
2. What physiological or social factors could lead to heat illness in the elderly?
3. How can drugs put elderly persons at risk for thermoregulatory disorders?
4. What is the evidence supporting the view that the elderly have an increased incidence of thermoregulatory impairment?
5. List specific neurological disorders which can lead to impairment of thermoregulation.
6. How can Alzheimer's disease be a cause of either hypothermia or heat illness?
7. Summarize the evidence from animal experiments for an influence of age on thermoregulation.
8. How can alcohol predispose to impaired thermoregulation?

PART 2
PATHOPHYSIOLOGY

Chapter 10

Learning Objectives
1. Understand the overall mechanism underlying the pathogenesis of infectious fevers.
2. Identify pathogenic factors thought to be implicated in infectious fever.
3. Categorize fevers of other than infectious origin.
4. Identify the currently known putative endogenous pyrogens.
5. Describe the possible sites and modes of action of peripheral cytokines.
6. Explain the postulated central neural and chemical control of fever production and lysis.
7. Know the thermoeffector mechanisms responsible for fever production and lysis.
8. Know the effects of fever on the body's functions.
9. Describe the manifestations of infectious fevers, *viz.*, onset, character, course, and duration.
10. Recognize the influence of age, size, time of day, ambient temperature, muscular activity, hypoxia, and other physical factors on the febrile response.
11. Discuss the biological significance of fever.

Bullets
- B-1 Fever and hyperthermia are distinct.
- B-2 Invading pathogens *per se* do not cause fever.
- B-3 Cytokines as endogenous pyrogens.
- B-4 The ultimate target of pyrogens is the POA.
- B-5 The means by which peripheral cytokines inform the POA are still controversial.
- B-6 PGE_2, a putative central fever mediator.
- B-7 The thermoeffector responses to pyrogens and to cold exposure are similar.
- B-8 Is fever good or bad?

Chapter 10

FEVER

CLARK M. BLATTEIS
Department of Physiology and Biophysics
The University of Tennessee, Memphis 38163, USA

1. Introduction

The term **fever** specifically defines the elevation of T_c characteristically exhibited by most species in response to their invasion by infectious agents, live or inanimate. It is the most manifest among an array of non-specific systemic reactions designed to combat the deleterious effects of invading pathogens, *i.e.*, to restore health to the afflicted host. These reactions are collectively termed the *acute-phase reaction* (APR) and constitute, therefore, the primary host defense response.

Hence, "fever" is distinct from "hyperthermia", and the two terms should not be used interchangeably. The particular difference to note is that, during hyperthermia, the T_c rise is the unavoidable consequence of the passive gain of heat in excess of the capability of active thermolytic effectors to dissipate it (see Chap. 11.1), whereas during fever, the T_c rise is the deliberate result of the active operation of thermogenic effectors. An effect of this difference is that febrile subjects prefer warm thermal environments while hyperthermic subjects choose cool ones, to facilitate heat storage and to enhance heat loss, respectively. Another important distinction between fever and hyperthermia is that, whereas the latter is clearly dependent on T_a, the former is independent of it, *i.e.*, fever can develop at any T_a. A practical lesson from this differentiation is that the correct diagnosis of the cause of an elevated T_c is important to determining the appropriate therapy.

B-1

2. Pathogenesis of Fever
2.1. The pyrogenic materials

Many different substances are capable of causing fever (Table 1). To the extent that they originate outside the body, they are called **exogenous pyrogens**. Mostly, they are infectious microorganisms or their products recognized as foreign by the invaded host's immune cells. However, fevers can also be due to noninfectious causes. The causative factors in these cases are host-derived, *e.g.*, immune complexes, allergens, nonmicrobial inflammation. Although these agents are not properly exogenous, they nevertheless are technically regarded as exogenous pyrogens. This is so because it is now well established that fever and its nonthermal acute-phase correlates are not induced directly by these original pathogenic materials, but rather by certain endogenous mediators, called **endogenous pyrogens**, generated in consequence of their encounter with the host's immune cells. Psychological stress may also be febrigenic in certain cases, but the T_c rise under these conditions is

B-2

probably mediated by one or more of the neuroregulators presumptively involved distal to the endogenous pyrogens (*e.g.*, prostaglandin E_2 [PGE_2]; see later).

Table 1. SOME COMMON PATHOGENIC STIMULI THAT INDUCE PYROGENIC CYTOKINES

A. **Microbial** (pyrogenic agents)
  Viruses (whole organism; hemagglutinin; dsRNA)
  Bacteria
    Gram-positive (whole organism; peptidoglycans [muramyl
      dipeptide], lipoteichoic acids, exotoxins, enterotoxins;
      erythrogenic toxins; group B polysaccharides)
    Gram-negative (whole organism; peptidoglycans;
      lipopolysaccharides [lipid A])
  Mycobacteria (whole organism; peptidoglycans; polysaccharides;
    lipoarabinomannan)
  Fungi (whole yeasts; capsular polysaccharides; proteins)

B. **Nonmicrobial** (pyrogenic agents)
  Antigens (*e.g.*, bovine or human serum albumin, bovine gamma globulin,
    ovalbumin, penicillin)
  Inflammatory agents (*e.g.*, asbestos, silicia, uv radiation, turpentine)
  Plant lectins (*e.g.*, concanavalin A, phytohemagglutinin)
  Drugs (e.g., polynucleotides [*e.g.*, polyinosinic:polycytidylic acid],
    antitumor agents [*e.g.*, bleomycin] plant alkaloids [*e.g.*, colchicine,
    vinblastine] synthetic immunoadjuvants [*e.g.*, muramyl peptides]
  Host-derived (*e.g.*, antigen-antibody complexes, activated complement
    fragments, inflammatory bile acids, urate crystals, certain androgenic
    steroid metabolites [*e.g.*, etiocholanolone], certain lymphocyte
    products)

2.2. Endogenous pyrogens

These factors (Table 2) belong to the class of immunoregulatory polypeptides called *cytokines*. Most prominent among these are interleukins (IL)-1β and -6, tumor necrosis factor (TNF)-α, and interferon (IFN)-α. They are produced by various cell types, but, at the outset, primarily by mononuclear phagocytes activated by various signals provided by the exogenous pyrogens (Table 3). Their production is complex, involving both self-amplification and the concatenated production and release of other cytokines. For example, following the administration of the bacterial product, lipopolysaccharide (LPS, a portion of the outer wall of Gram-negative bacteria [see Table 1]), TNF-α normally appears first, followed by IL-1β, and finally IL-6, the

B-3

release of the latter two being stimulated by the first, and that of the last also by the second (see Table 2). However, IL-6 does not induce TNF-α and IL-1β; to the contrary, it suppresses their expression. In addition, these cytokines auto-regulate

Table 2. PUTATIVE INTRINSICALLY PYROGENIC CYTOKINES

Tumor Necrosis Factors (TNF)	Interleukins (IL)	Interferons (IFN)	GP130 Ligand	Other Cytokines
$\alpha^{1,9}$ (cachectin)	$1\alpha^{1,2,9}$	$\alpha^{6,8}$	IL-$6^{3,8}$	Granulocyte-Macrophage Colony Stimulating Factor
$\beta^{4,5}$ (lymphotoxin)	$1\beta^{1,2,9}$	β^3	IL-11	(GM-CSF) ^{4,5}
	$2^{3,4,5}$	$\gamma^{1,4,6,8}$	Ciliary Neurotrophic Factor (CNTF)	Macrophage Inflammatory Protein-1α/β (MIP-1α/β) ^{3}
	$8^{3,4}$		Leukemia Inhibitory Factor (LIF)	Acidic Fibroblast Growth Factor (aFGF)
	$12^{4,7}$		Oncostatin M (OM)	
			Cardiotropin-1 (CT-1)	

Some interactions among the principal endogenous pyrogens: [1]Induced by each other; [2]inducible by itself; [3] induced by IL-1 and/or TNF-α; [4] intrinsic pyrogenicity is controversial; [5]induces IL-1, TNF-α, and/or IFNγ; [6]enhances LPS- and TNF-α-induced IL-1; [7]induced by IL-1, IL-6, and/or TNF-α; [8]suppresses IL-1-induced TNF-α and IL-1 production; [9] Induced by GM-CSF.

their levels through the release of their own functional antagonists, *e.g.*, specific cell-surface receptor antagonists and soluble receptors, and/or of inhibitors of their synthesis. Table 4 lists various endogenous substances that may thus modulate peripherally the pyrogenic action of different cytokines. It has consequently been difficult to discern which cytokine is the critical fever-producing agent. Indeed, the regularity of the cascade, TNF-α→ IL-1β→ IL-6, initiated by LPS, has lately been questioned because, *e.g.*, normal fevers develop in mice lacking TNF-α and IL-1β receptors, but are impaired in IL-6-deficient mice following intraperitoneal injections of LPS, suggesting that IL-6 rather than TNF-α and/or IL-1β may be the primary pyrogenic cytokine. This may be relevant because, in humans treated with LPS, blockade of IL-1 receptors with a specific antagonist, IL-1 receptor antagonist (IL-1ra), does not prevent fever. But, on the other hand, C3H/HeJ mice, which can not produce TNF, also do not develop fever to LPS, although they do to exogenous TNF-α.

Table 3. POTENTIAL CELL SOURCES OF CYTOKINES

Hemopoietic Stem Cell-Derived

Myeloid lineage
Mononuclear phagocytes
Monocytes
Macrophages (hepatic [Kupffer cells], alveolar, glomerular, mesangial, serosal, peritoneal, splenic, lymph node sinus, microglial, astrocytic, synovial, placental; neoplastic cell lines)
Polymorphonuclear granulocytes (neutrophils, eosinophils, basophils, mast cells)
Platelets
Lymphoid lineage
Lymphocytes (T-, B-, and NK cells; neoplastic cell lines)

Other Cells
Epithelial cells (keratinocytes, Langerhans cells, corneal, gingival, thymic)
Fibroblasts; chondrocytes; osteoblasts; neurons
Endothelial, vascular smooth muscle, uterine stromal, anterior pituitary, dendritic, renal mesangial cells
Others

2.3. Afferent signaling

Since exogenous pyrogens most commonly enter the body through a break in the skin or through the respiratory, digestive, or urogenital system, their activation of phagocytic cells and the resultant release of endogenous pyrogens are presumed to occur peripherally also. Since, furthermore, T_c is regulated in the POA (see Chap. 6), it is generally thought that the cytokines released by these cells are transported by the bloodstream to this brain region, which they then activate. Indeed, they readily induce fever when microinjected intracerebrally. There is no evidence that pyrogenic cytokines have a direct action on either thermoafferent or thermoefferent structures *per se*. In the POA, they inhibit the activity of warm-sensitive and enhance that of cold-sensitive neurons, consistent with the diminished heat loss and enhanced heat production that these outputs, respectively, modulate, thus supporting the notion that

B-4

Table 4. SOME ENDOGENOUS SUBSTANCES WITH POSSIBLE PERIPHERAL MODULATORY ROLES IN THE FEBRILE RESPONSE TO DIFFERENT CYTOKINES

Mediators	Cytokines reportedly affected
A. Facilitatory (pyretic)	
1) Upregulate expression	
Tachykinins (e.g., SP)[1]	TNF-α, IL-1β, IL-6
Complement (C3a, C5a, MAC) [1,2]	TNF-α, IL-1β, IL-6, PGE$_2$
Prostaglandin E$_2$ (PGE$_2$)	IL-1α, IL-6
Norepinephrine (NE)	IL-6
Serotonin (5-hydroxytryptamine, 5-HT)	IL-1β
B. Inhibitory (antipyretic)	
1) Downregulate expression	
Glucocorticoids[3], PGE$_2$, NE,	TNF-α, IL-1β, IL-2, IL-6,
lipoxygenase inhibitors	IL-8, IL-12, IFNα, β, γ, GM-CSF
IL-4, IL-6, IL-10, IL-13,	
IFN-γ, TGF-β[4]	TNF-α, IL-1β, IL-6
Nitric oxide (NO) [5], O$_2$ radicals	IL-6
Heat shock proteins (HSPs)	TNF-α, IL-1β
2) Bind receptors	
IL-1 receptor antagonist (IL-1ra)	IL-1α, β
Soluble receptors	TNF-α, IL-1α, β
(Non-target) cell surface receptors	Various
Natural receptor fragments	Various
Protein carriers (e.g., α$_2$-macroglobulin)	IL-1β, IL-6
3) Attenuate action	
Lipocortin-1	IL-1β, IL-6, IL-8, IFN-γ
αMSH	Various
TNF-α[6]	Various

[1]Also induce PGE$_2$; [2]elicited only by certain ExP (e.g., LPS, Ag-Ab complexes); [3]also suppress phospholipase A$_2$ and cyclooxygenase-2; expression may stimulate αMSH induction; increase receptor expression of IL-1β, IL-6, IFNγ, and GM-CSF (but not affinity); [4]TGF-β also promotes IL-1ra synthesis; [5]also potent activator of cyclooxygenase-1 and -2; [6] inhibits peripherally LPS-induced fevers only.

thermosensitive units may be their direct targets. Moreover, the central administration of their antagonists suppresses fever. However, it is controversial how circulating cytokines reach the POA, since, as large hydrophilic peptides, they are unlikely to cross the blood-brain barrier (BBB) by passive diffusion. While some data [1] indicate that these cytokines may be actively transported across the BBB into the intact brain, the time course and quantity of this transport are too slow and minimal, respectively, to account for the rapid onset of fever following, especially, their intravenous (iv) administration. Other evidence [2] suggest that, alternatively, they may preferentially interact with sensory elements in or near circumventricular organs, brain structures that lack a BBB; the *organum vasculosum laminae terminalis* (OVLT), on the midline of the POA, may be such a structure. This interaction, in turn, is thought to evoke secondary, neural and/or chemical signals that transduce the blood-borne cytokines' pyrogenic messages inwardly to the POA. It is unclear, however, which of the various cell types in the OVLT region that express cytokine receptors actually provide the signals specifically for fever production. Because of their strategic location at the interface between blood and brain, it has been suggested [3] that the luminal side of cerebromicrovascular endothelial cells, in particular, may be the primary target of circulating endogenous pyrogens; endothelial cells release IL-1β upon LPS and IL-1 activation (see Table 2). The thus-induced cytokines are postulated then to be secreted abluminally and, by paracrine actions, to affect the activities of local thermosensitive neurons, and/or to stimulate microglial cells in the region to produce additional cytokines; the latter would amplify and prolong the original signals. Indeed, there is little doubt that TNF-α, IL-1β, and IL-6 are ultimately induced in brain by systemic pyrogens. However, the kinetics of the various synthetic processes involved are too slow to account for the released substances having roles in the *initiation* of the febrile response to circulating pyrogens, although they could play roles in its subsequent maintenance. Hence, to substantiate the short latency of the febrile response to iv-injected cytokines, a more rapid means of signaling between peripheral endogenous pyrogens and the POA may be presumed to exist. Recently, sensory vagal afferent nerves distributed in the vicinity of the presumptive, primary source of cytokine production, *i.e.*, hepatic macrophages (Kupffer cells), have been implicated in LPS and IL-1β fever genesis. Neural afferents (C fibers) from other sites, *e.g.*, abdomen, have also been implicated. Thus, evidence has been adduced that these terminals bind cytokines, and that the neural message is conveyed to the nucleus tractus solitarius, the primary medullary projection area of the vagus nerve, then passes to the POA, probably via ascending noradrenergic projections originating in the A1/A2 cell groups of the medulla oblongata; norepinephrine (NE) levels rise promptly in the POA after a peripheral pyrogenic challenge. The exact nature of the transducing signals in the POA has not yet been elucidated, but NE is an agonist of phospholipases C and D, enzymes that degrade membrane phospholipids yielding arachidonic acid and, ultimately, PGE_2. This latter mediator causes a prompt elevation of T_c when microinjected directly into the POA, and its level and receptors in the POA region increase following peripheral LPS or cytokine administration, in correspondence with the febrile rise [4]. Moreover, PGE_2 synthesis inhibitors (*e.g.*, aspirin) are potent antipyretics. Nevertheless, although the notion that PGE_2 is an obligatory central mediator of fever is widespread, certain cytokines (*e.g.*, MIP-1α) reportedly evoke rapid-onset fevers that are unaffected by prior administration of PGE_2 synthesis inhibitors. Finally, fever can occur even in the absence or deafferentation of the POA, indicating that subsidiary sites also exist that both express sensitivity to pyrogenic mediators (cytokines, PGE_2, NE) and can modulate the febrile response. Such putative sites have been localized in the hypothalamic paraventricular nucleus and the medulla oblongata, among others [5]. In sum, the mechanism by which peripheral pyrogenic signals are translated into brain signals that modulate fever remains to be elucidated.

B-6

3. Pathophysiology of Fever
3.1. Resetting of the set-point

Following control system theory (see Chap. 3), fever is regarded as a regulated, upward adjustment of the thermoregulatory set-point, induced by a pyrogen. This means that the elevated T_c in disease is maintained and, importantly, defended as efficiently as its cenothermic level in health; *i.e.*, as already mentioned, exposure to cold or heat (or physical exercise) during steady-state fever evokes heat production and loss responses, respectively, not different than those in cenothermy. When the pyrogenic stimulus has abated, the set-point and, consequently, T_c return to normal. The central mechanisms by which the control of T_c is thus altered are still unknown.

Various putative neuroregulators, both facilitatory and inhibitory, have been implicated, but it is not yet clear how these factors are organized and integrated in the neuromodulation of the febrile response. Table 5 lists various endogenous factors that have presumptive roles in the central modulation of fever. (Be aware that Tables 4 and 5 are not inclusive and, in particular, that the modes of action shown are tentative and controversial.)

3.2. The course and character of fever

Following the introduction of a pyrogen into the body, there is an interval during which the various host-generated reactions that eventuate in the development of fever take place; T_c is not yet changing. This period is called the *latent* or *prodromal period*. Its duration depends on the nature of the exogenous or endogenous pyrogen, the route of its administration, its amount, etc. It is succeeded by the *phase of rising* T_c. The height, duration, and other characteristics of the ensuing fever are also largely dependent on the amount and type of the original pyrogen. They can, however, be modified, like the cenothermic T_c (see Chap. 2), by various endogenous (*e.g.*,

Table 5. SOME ENDOGENOUS SUBSTANCES WITH POSTULATED CENTRAL MODULATORY ROLES IN THE FEBRILE RESPONSE TO DIFFERENT CYTOKINES.

Mediators	Implicated in fever caused by	Possible actions
A. *Facilitatory (pyretic)*		
Prostaglandin E_2 (PGE$_2$)	TNF-α, IL-1β, IL-6, IFN-α, CNTF	Putative second messenger; downregulates warm-sensitive neurons
β-endorphin (β-E)	TNF-α, IL-1β, IL-6, IFN-α	Downregulates warm-sensitive neurons
Corticotropin-releasing h. (CRH)	IL-1β, IL-6, IL-8, IFN-γ (not IL-1α, TNF-α)	May activate effector neurons, after PGF$_{2\alpha}$ (not PGE$_2$); dose- and species-dependent
Cholecystokinin-8 (CCK-8)	(1)	Induces PGE$_2$
Substance P (SP) (1)		Induces TNFα, IL-1β, PGE$_2$, CCK-8
Norepinephrine (NE)	(1)	Induces PGE$_2$, CRH, peripheral IL-6
B. *Inhibitory (antipyretic)*		
Arginine vasopressin (AVP)	(1)	In ventral septum
α-Melanocyte-stimulating h. (αMSH)	(1)	Mechanism uncertain
Lipocortin-1 (LP-1)	IL-1β, IL-6, IL-8, IFN-γ	Inhibits CRH and arachidonic acid (hence, PGE$_2$) synthesis
Serotonin (5-hydroxy-tryptamine, 5HT)	(1)	
γ-Aminobutyric acid (GABA)	(1)	Inhibits POA warm-sensitive neurons

(1) Based on response to ExP, crude EnP, one cytokine (usually IL-1β), or PGE$_2$.

hydrational status) and exogenous (*e.g.*, climatic) factors; for example, fever is often augmented during dehydration and at high T_a (see below). When T_c reaches its maximum, it remains there for a period of time, termed the *stable* or *plateau phase* or *fastigium*. Its duration is also related to the dose of the pyrogen, but can be influenced by extraneous factors. Finally, the fever breaks ("crisis"), and T_c begins its return toward the cenothermic level. This phase is variously called the *phase of falling T_c*, *fever lysis*, or *defervescence*.

3.3. Thermoeffector mechanisms of fever production and lysis

Functionally, the upward shift in the set-point of T_c is expressed by an increase in metabolic heat production and cutaneous vasoconstriction, and by cessation of sweating, if present; *i.e.*, the responses evoked are analogous to those caused by acute cold exposure. The former is associated with the most visible sign of fever, shivering ("chills"), while the latter is manifested by a cold and pale skin, provoking subjective sensations of cold that prompt the febrile subject to actively (*i.e.*, behaviorally) seek warmer surroundings. During the plateau phase of fever, the cutaneous vasculature resumes its normal, relatively constricted state (inappropriate as compared to an expected vasodilated state at a T_c comparably elevated by heat exposure, but thereby helping to maintain T_c febrile). Since the blood now perfusing the skin is warm, T_{sk} rises and the skin condition changes to warm and pink; consequently, the earlier sensation of cold disappears. The relative contributions of decreased heat loss and increased heat production vary, depending on the pyrogenic dose and/or the T_a; *i.e.*, an attack of fever in the cold may require intense heat production, whereas in the heat a decrease in heat loss may suffice. In very high T_a, the added passive heat gain from the environment can cause T_c to become dangerously high; dehydration compounds the danger. When eventually the set-point of T_c returns to its cenothermic level, the effector mechanisms evoked resemble those in heat-exposed subjects, *viz.*, T_c falls in conjunction with cutaneous vasodilation and drenching sweating, and the defervescing subject seeks a cooler environment.

3.4. Physiological support of fever

The febrile rise is associated with various circulatory and respiratory adjustments that serve to support the increased metabolic demands of thermogenic tissues. These include increases in heart rate and cardiac output, increased blood flow to the heat-producing organs (*e.g.*, muscle, fat), and a transient hyperventilation resulting in a fall in $P_{a_{CO2}}$ and a rise in pH_a; $P_{a_{O2}}$ is generally unaffected. Other changes involve endocrine, enzymatic, and cellular effectors associated with the provision and utilization of energy for sustaining the enhanced heat production, analogous to cold exposure.

4. Fever as a Clinical Sign

The duration, pattern, and magnitude of natural fevers vary, the fever intensity most generally being related to the type, but inconsistently with the severity, of the infection [6, 7]. In humans in a thermoneutral environment, the febrile rise typically ranges from 0.5 to 3 °C; thus, most infectious diseases produce fevers between 38 °C (100.4 °F) and 40.5 °C (104.9 °F). An average fever is 39.5 °C (103.1 °F); there are no statistically significant gender or racial differences. Extreme pyrexia, $i.e.$, T_cs above 42 °C (107.6 °F), is a rare occurrence; various mechanisms have been implicated in setting this upper limit [8; see also Tables 4 and 5]. On the other hand, some infected patients remain afebrile or become hypothermic. The latter is usually a bad prognosis. It is a shock-associated phenomenon, manifested in severe sepsis and attributed to a decrease in the threshold T_c at which thermogenic effectors are activated. This results in a widening of the normally narrow interthreshold zone between heat- and cold-defense mechanisms; T_c, therefore, falls until the threshold for thermogenesis is reached. Interestingly, this is accompanied by a marked preference for a cooler T_a, presumably to hasten the T_c fall to its new threshold [9, 10].

The occurrence of fever is not a consistent event under all conditions. For example, febrile sensitivity is attenuated in neonates and the elderly. Responsiveness to pyrogens is also reduced during night-time sleep, at high terrestrial altitudes, at term of pregnancy, in malnutrition, and by certain forms of stress. Serious infection may consequently go undetected until the appearance of other signs, putting patients under such conditions at greater risk.

Fevers have been classically described by their patterns as intermittent, remittent, sustained, and relapsing. However, recent studies have indicated that these descriptions are not, in fact, characteristic of any infectious disease or group of diseases, so that these signs, even in the absence of a definite diagnosis by other means, are no longer considered to be determinate.

5. Benefits and Risks of Fever

It should be apparent from the preceding that, contrary to how it is sometimes described, fever is *not* a disorder of temperature regulation, but rather a functional elevation of T_c that is an integral part of coherent, organized, systemic, homeostatic responses serving to resist the potentially deleterious effects of pathogenic agents, $i.e.$, the APR. Therefore, it may be presumed that fever is beneficial. Indeed, elevations of temperature that mimic those in fever have been reported in some *in vivo* and *in vitro* experimental models of infection to enhance various immune functions associated with increased resistance to infection. Its demonstrated benefits include the enhancement of phagocytosis, neutrophil migration, T-cell proliferation and O_2 radical production, the increased synthesis of IFN and the augmented antiviral and antitumor activities of this cytokine, the decreased growth rate and viability of iron-dependent bacteria, and, presumably, improved host survival rate since, as already mentioned, the lack of fever during infection may be a bad prognosis. There is little evidence, however, that a T_c rise *per se*, *i.e.*, without other, associated acute-phase components, is beneficial, probably because clinical fever is not an isolated event, but one among the various, interrelated, not easily separated, systemic host defense responses that together characterize the APR. Hence, although antipyretic therapy is widely practiced without apparent untoward effect on the outcome of infection, it would nevertheless seem advantageous to let the fever run its course, especially if there is no impending threat to vulnerable organs, such as the brain, from extreme elevations of T_c or other, relevant contraindications.

B-8

But high or prolonged fevers may potentially also have harmful effects. These include dehydration, delirium, localized lesions (*e.g.*, liver, brain), convulsions, cardiopulmonary strain, negative nutrient balance, and possible teratological consequences (particularly micrencephaly if maternal fever > 2.5 °C and persists for 8 days or more during early pregnancy). The increased metabolic rate of febrigenesis may also represent a potential strain for those whose energy reserves may be limited, such as the malnourished, elderly, neonates, cancer patients, and patients with metabolic diseases. Similarly, the associated tachycardia and polypnea may put at risk patients with a diminished capacity to increase cardiac work (*e.g.*, congestive heart failure) or to hyperventilate (*e.g.*, chronic obstructive pulmonary disease, asthma, silicosis, etc.). Patients with dysfunctional kidneys are also imperiled. Between 6 months and 3 1/2 years of age, some febrile children are susceptible to the development of ectopic foci of CNS activity that are expressed as brief, epileptic-like seizures (*infantile febrile convulsions*); however, these are usually without consequence or recurrence.

6. Summary

Fever is the most manifest of an array of nonspecific host defense responses to infection and inflammation, collectively called the acute-phase reaction. The term "fever" specifically defines the regulated, *active* rise in T_c induced by pyrogens, different than the passive T_c rise consequent to heat gain by environmental exposure or physical activity. It is now generally believed that fever is generated not by exogenous pyrogens (*i.e.*, the original, invading pathogenic microorganisms and their products), but by endogenous pyrogens released, predominantly, by the host's mononuclear phagocytes activated by the invading exogenous pyrogens.

Endogenous pyrogens are members of the class of immunoregulatory polypeptide mediators called *cytokines*. Their pyrogenic action is presumably exerted in the POA, the primary locus of the neurons modulating thermoregulatory responses. Indeed, cytokines, whether injected peripherally or centrally, equally change the firing rates of warm- and cold-sensitive neurons in a manner consistent with the decreased heat loss and increased heat production that underlie the development of fever. It is controversial, however, how cytokines, as hydrophilic peptides, may interact with neurons in the POA if their free passage into the brain is excluded by the presence of the BBB. Various mechanisms have been postulated. Recent data suggest that fever may be initiated peripherally by as yet indeterminate mediators that activate neural afferents. In the brain, these inputs may be transmitted to the POA via noradrenergic pathways; PGE_2 may be the ultimate central mediator. Various other neuroregulators appear to have roles in fever genesis and lysis, but it is not yet clear how their actions are integrated in the overall neuromodulation of the febrile response.

Clinical fevers exhibit different patterns, durations, and magnitudes, related, most generally, to the type of the infection. In addition, a variety of other factors may influence the course and character of infectious fevers. Unless there is a failure of thermoregulatory mechanisms, however, a pyretic maximum exists (*ca.* 42 °C) that is almost never exceeded. This upper limit may be set, in part, by the antagonistic actions of various endogenous neurochemicals elicited by the pyrogens or their secondary products.

Fever, as a host defense response during infections, presumptively benefits the host. Indeed, various immune functions associated with resistance to infection are enhanced by induced fever-like T_c rises. Hence, it may be advisable not to treat the febrile rise *per se* if there is no threat from the potentially harmful effects of high or prolonged fever or in the absence of high-risk predisposing factors.

7. References

[1] Banks WA, Kastin AJ, Broadwell RD (1996). Passage of cytokines across the blood-brain barrier. *Neuroimmunomodulation* 2: 241-248.

[2] Blatteis CM, Sehic E (1997). Fever: How may circulating pyrogens signal the brain? *News Physiol. Sci.* 12: 1-9.

[3] VanDam A-M, deVries HE, Kuiper J, Zijlstra FJ, deBoer AG, Tilders FJH, Berkenbosch F (1996). Interleukin-1 receptors on rat brain endothelial cells: a role in neuroimmune interaction? *FASEB J.* 10: 351-356.

[4] Blatteis CM, Sehic E (1997). Prostaglandin E_2: A putative fever mediator. In: Mackowiak PA (ed), *Fever: Basic mechanisms and management*, 2nd ed. Lippincott-Raven, Philadelphia, pp. 117-145.

[5] Blatteis CM (1992). The pyrogenic action of cytokines. In: Rothwell NJ, Dantzer RS (eds), *Interleukin-1 in the brain*. Pergamon, Oxford, pp. 93-114.

[6] Atkins E, Bodel P (1979). Clinical fever: its history, manifestations and pathogenesis. *Federation Proc.* 38: 57-63.

[7] Mackowiak PA (1997). Relationship between the height of a fever and severity of the illness. In: Mackowiak PA (ed), *Fever: Basic mechanisms and management*, 2nd ed. Lippincott-Raven, Philadelphia, pp. 251-254.

[8] Mackowiak PA, Boulant JA (1997). Fever's upper limit. In: Mackowiak PA (ed), *Fever: Basic mechanisms and management*, 2nd ed. Lippincott-Raven, Philadelphia, pp. 147-163.

[9] Romanovsky AA, Shido O, Sakurada S, Sugimoto N, Nagasaka T (1996). Endotoxin shock: Thermoregulatory mechanisms. *Am. J. Physiol.* 270: R693-R703.

[10] Romanovsky AA, Kulchitsky VA, Akulich NV, Koulchitsky SV, Simons CT, Sessler DI, Gourine VN (1996). First and second phases of biphasic fever: two sequential stages of the sickness syndrome? *Am. J. Physiol.* 271: R244-R253.

Additional suggested reading:

Bartfai T, Ottoson D (eds) (1992). *NeuroImmunology of fever*. Pergamon, Oxford.

Cunha BA (ed) (1996). Fever. *Infect. Dis. Clin. N. Am.* 10: 1-222.

Kluger MJ (1991). Fever. Role of pyrogens and cryogens. *Physiol. Rev.* 71: 93-127.

Mackowiak PA (ed) (1997). *Fever: Basic Mechanisms and Management*, 2nd ed. Lippincott-Raven, Philadelphia.

Milton AS (ed) (1982). *Handbook of Experimental Pharmacology, vol. 60. Pyretics and antipyretics.* Springer, Berlin.

Toth LA, Blatteis CM (1996). Adaptation to the microbial environment. In: Fregly MJ, Blatteis CM (eds), *Handbook of Physiology, Sec. 4: Environmental Physiology, vol. II.* Oxford Univ. Press, New York, pp. 1489-1519.

Self-study questions
1. Are fever and hyperthermia synonymous?
2. Distinguish between exogenous and endogenous pyrogens.
3. Where in the brain do the pyrogenic cytokines act?
4. How may peripheral cytokines signal the POA?
5. What are the effects of cytokines on preoptic thermosensitive neurons?
6. Which second messengers may mediate the pyrogenic action of cytokines?
7. Which endogenous factors may determine the upper limit of fever?
8. Name the stages of fever.
9. Describe the thermoeffector mechanisms of fever genesis and lysis
10. What are some factors that can influence the febrile response?
11. What is the function of fever in the context of the acute-phase reaction?
12. Should fever be treated?

Chapter 11

Learning Objectives
1. Basic understanding of the integrative acclimatory response: sweating, cardiovascular system and metabolic changes.
2. Body fluid compartmentalization in the acclimated state.
3. Cellular components of heat acclimation: signal transduction pathways, hormonal effects and the role of heat shock proteins.
4. Understanding of the dynamics of heat acclimation.
5. Loss of acclimation and its re-induction.
6. Problems concerned with human acclimation.

Bullets

B-1 The classic description of heat acclimation response constitutes lowered internal core temperature, decreased heat production, lowered heart rate . . .
B-2 Adaptation of the heat dissipation effector-organs.
B-3 Rerouting of evaporating water during heat stress and thermal dehydration.
B-4 There is substantial evidence that a variety of membranal and cellular events to improve cellular performance take place.
B-5 Does expanded optimal temperature margins via increased stock HSP level occur?
B-6 Heat acclimation: A bi-phasic model of central-peripheral cross talk.
B-7 How fast acclimation is lost?
B-8 Despite the variety of species and taxonomic groups, the basic principle of heat acclimation is similar.

Chapter 11.1

PROLONGED EXPOSURE TO HEAT

MICHAL HOROWITZ
*The Hebrew University, Hadassah Medical School
Department of Physiology, 91120 Jerusalem, Israel*

1. Introduction

Animals, as well as humans, display an ability to adjust their physiological mechanisms to overcome long-term shifts in ambient temperatures. Persistent heat stress induces a variety of changes at all levels of body organization, ultimately leading to decreased strain and increased thermal endurance and/or an improvement in tolerance to a higher core temperature. It is now clear that both intrinsic and extrinsic (e.g., intracellular vs. extracellular) adjustments play key roles in the adaptation process. Depending on the adaptive requirements, the emerging changes either increase or decrease the expression of a variety of features. Opposite responses, which produce adaptive "conflicts", may also lead to the exclusion of features from the adaptive repertoire. Thus, although adaptations lead to better performance under further stress, they sometimes involve compromises.

The process of adaptation occurring under laboratory-induced conditions is called acclimation; when it occurs under natural conditions, it is called acclimatization. In general, adaptations emerging upon short term acclimatization (e.g., seasonal climatic variations) are similar in nature to those conferred by acclimation, although the magnitude of the response may differ, depending on the severity of the climatic conditions and other environmental cues which are sometimes influential (e.g., photoperiod). The adaptations found in both heat acclimation and seasonal acclimatization may differ from those evolved in species inhabiting extremely hot climates, although acclimation, to a certain extent, is a recapitulation of the evolutionary adaptive process. It is a highly complex process, with physiological adjustments occurring at all levels of body organization.

The major part of this chapter deals with problems concerned with heat acclimation of the thermoregulatory effectors in mammalian species. Comparative aspects of heat acclimation/acclimatization are dealt with in a separate section, at the end.

2. Integrative Acclimatory Responses of the Thermoregulatory Effectors in Homeotherms

2.1. Global acclimation response

The classic description of the heat acclimation response in mammalian species constitutes lowered internal body temperature, decreased heat production, lowered heart rate, elevated stroke volume, augmented thermoregulatory skin blood flow, and increased capacity of the evaporative cooling system. Altogether, these features are the outcome of integrated central and peripheral modulations. Decreased temperature thresholds for activation of vasodilatation and evaporative cooling, and increased thermal sensitivity of the control center, together with greater capacity of the peripheral heat dissipating effectors lead to enhanced resistance during heat stress. Concomitantly, metabolic heat production, both under sedentary conditions and during work, decreases, reducing the internal strain. Increased endurance of 60-70% is gained via these mechanisms.

2.2. Global adaptation of the evaporative cooling system

The capacity of the evaporative cooling system increases in all species following prolonged exposure to moderate heat. Nevertheless, there are species variations, depending on the cooling mechanism. Among mammalian species, extensive studies on acclimation of this system have been carried out only in humans and primates (sweating) and rodents (salivation). With both mechanisms, cooling power is enhanced by a lowering of the temperature threshold for water secretion, allowing a widening of the activity range, by a greater rate and volume of secretion, and by augmentation of the wetted surface area. Concomitantly, the energy cost of secretion is significantly lower than in the preacclimation state due to increased efficiency of the secretory mechanisms. It is noteworthy, however, that while sweating species augment sweat volume with increased temperature, acclimation in rodents is directed towards water conservation, and increased capacity of the salivary gland is manifested only upon severe stress.

2.3. Adaptation of the cardiovascular system

The global adaptation of this system is manifested by an increase in the cardiovascular reserve to improve the capacity for heat transfer upon heat stress. This is brought about by commensurate increases in both cardiac and peripheral vascular capacities which, together, improve stroke volume and splanchnic and skin blood flows.

2.3.a. Peripheral blood flow

Greater skin blood flow is achieved either by widening the dynamic range via increasing basal skin vascular tone during normothermic conditions (e.g., humans, sheep) or by decreasing basal tone without much elevation of blood flow upon heat stress (e.g., rodents). Heat acclimation also increases the share of cardiac output distributed to the splanchnic organs. Thus, in the acclimated subject, greater heat-induced splanchnic vasoconstriction allows greater thermoregulatory blood flow through cutaneous vascular beds, coincidentally with the maintenance of adequate splanchnic blood and heat flow from internal organs (Fig. 1).

2.3.b Chronotropic response of the heart and the regulation of stroke volume

Following heat acclimation, increased peripheral thermoregulatory vascular demands are met by a concerted array of peripheral mechanisms and intrinsic cardiac adaptations which, collectively, increase the cardiac reserve. Adaptation of the heart is manifested by decreased heart rate and increased stroke volume. In humans, the decrease in heart rate has been attributed to lowered internal body temperature. However, recent animal studies suggest that changes in autonomic outflow and intrinsic changes in the atrial pacing cells both contribute to heat acclimation-induced bradycardia. The contribution of each component in the control of heart rate shows species variation. In the rat, for example, there are changes in the intrinsic heart rate, increased tonic activity of the parasympathetic

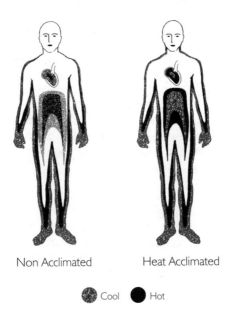

Figure 1. Cardiac output and regional blood flow before and during heat stress in non-acclimated and heat-acclimated subjects. It is noteworthy that heat-acclimated heat-stressed subjects show a remarkable increase in both stroke volume and splanchnic blood flow.

pathway, and slight withdrawal of sympathetic outflow. In contrast, in the sand rat *(Psammomys obesus)*, a diurnal desert species, unchanged vagal and sympathetic tone suggest that the major component contributing to heat acclimation-induced bradycardia is a rapid decrease in intrinsic heart rate. In this species, the contribution of the sympathetic outflow to the control of heart rate exceeds that of the parasympathetic outflow. Differences in the share of parasympathetic as compared to sympathetic innervation in heart rate regulation may thus stem from species variations. During heat stress, some species show unchanged cardiac output, whereas cardiac output is markedly elevated in others. In the hyperthermic rat, cardiac output is not elevated, unlike in the sand rat, which shows a marked elevation. This may imply that changes in sympathetic outflow distribution are responsible for the differences in heat-induced cardiac output response in these two species.

Increased stroke volume is the important adaptive feature of heat acclimation to increase cardiovascular reserve. Substantial data are now available suggesting that intrinsic changes in the myocardium and its contractile apparatus play a significant role in this important adaptation. The pressure generated by acclimated hearts is markedly greater than before acclimation. Concomitantly, increased left ventricular compliance allows the ventricle to deliver an increased stroke volume without much increase in end-diastolic volume, leading in turn to increased cardiac output with less energy expenditure. The increase in chamber compliance and cardiac contractility allows the heart to accommodate better to the increase in venous return, which, at least in part, is facilitated by the plasma volume expansion occurring upon heat acclimation (see *3.1*).

3. Body Fluid Compartments in the Acclimated State
3.1. Plasma volume expansion
Plasma volume expansion is often observed on heat acclimation. To some investigators, this feature provides the single, most important mechanism responsible for acclimation. However, not all investigations support its occurrence. The inconsistency in this response stems from variations in acclimation regimens (e.g., heating protocols, sedentary vs. active acclimation, and duration of the treatment; see *6.*). In contrast to the heat dissipation effectors, which show their full capacity when the complete acclimation state has been achieved (see *4*), plasma volume expansion *per se* plays its important role prior to this stage. The major mechanism for plasma volume expansion is an increase in total protein mass in the plasma compartment. This provides the osmotic power for water shift/retention into this compartment. Several mechanisms, such as altered protein extravasation, accelerated protein return via the lymphatic pathway (in the case of combined heat acclimation and exercise training), and increased plasma protein synthesis, operate in concert to induce this increase in plasma volume.

3.2. Compartmentalization of body fluids – rerouting of evaporative water during heat stress and thermal dehydration
Upon heat stress and thermal dehydration, a major manifestation of heat acclimation is the rerouting of the evaporative water from the plasma to the intracellular water compartment (Fig. 2). This is controlled by adjusting plasma protein levels in the circulating blood. These provide the osmotic force required for water retention in the plasma compartment. When subjected to thermal dehydration, the permeability of the vasculature to proteins is altered.

Heat-acclimated animals show decreased protein extravasation compared to non-acclimated animals. Thus, in acclimated animals, total plasma protein mass is elevated, favoring water retention in the plasma compartment. In non-acclimated animals under similar conditions, plasma protein mass and plasma volume decrease. Due to these differences, the share of the plasma compartment in the loss of body water by evaporation in heat-acclimated rodents is almost zero and most of the water lost by evaporative cooling comes from the intracellular compartment, thus conserving plasma water. In contrast, in non-acclimated rats, only 38% of the evaporated water originates intracellularly. It is noteworthy that the shift in water loss distribution in heat-acclimated rats is similar to that in desert rodents.

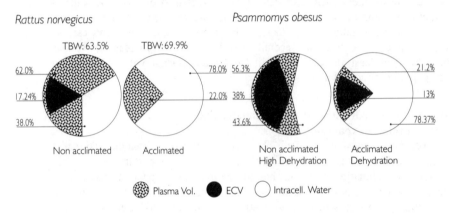

Figure 2. Distribution of total body water loss after dehydration in non-acclimated and heat-acclimated *Rattus norvegicus* and *Psammomys obsesus*. Upon acclimation, the contribution of intracellular water to evaporative water loss increases. Circle area denotes total water loss (100%). It is noteworthy that, before acclimation to heat, the share of intracellular water to evaporation is higher in the desert species. Redrawn from [5]

4. Energy Metabolism

Heat acclimation lowers the resting metabolic rate. Concomitantly, the lower critical temperature of the thermoneutral zone shows a significant shift to the right. When thermogenic capacity is measured by the rise in O_2 consumption, increased metabolism is sometimes more pronounced after heat acclimation, suggesting better energy reserves. This finding is somewhat controversial, however, since different groups report conflicting results. VO_2max during exercise in heat-acclimated subjects is also still under debate. While several groups did not find any acclimation effect, others reported a decreased VO_2max after heat acclimation. Nevertheless, the heat-acclimated body becomes more energetically efficient; hence, reduction in VO_2 max does not interfere with exercise intensity.

The decreased metabolic rate occurring upon heat acclimation is associated with the decrease in thyroid hormone levels known to occur with progression of heat acclimation, particularly when acclimation is undertaken under sedentary conditions. Significant interaction is also found between the effects of thyroid hormones and the acclimation temperature on obligatory thermogenesis. It is well documented that thyroid hormones are involved in metabolic shifts via their action on gene expression. This effect will be discussed further in 5.

A noteworthy energetic metabolic shift after acclimation leads to glycogen sparing and improved fat utilization during exercise. This enhances the muscles' buffering capacity and the ability to perform highly intense exercise following prolonged exertion in the heat. There is controversy regarding this finding and further investigation is required.

5. Cellular Mechanisms of Acclimation

There is substantial evidence that a variety of membranal and cellular events that improve the cellular performance of the thermoregulatory effectors are an integral part of the acclimation repertoire. The cellular adaptations can be categorized into those improving effector function and those increasing tolerance to higher core temperatures. Most of these adaptations are slow, being fully expressed only after several weeks of acclimation.

5.1. Thyroid hormones level and cellular work efficiency

Improved cellular work efficiency is associated primarily with energy metabolism. The metabolic adaptations discussed in 4 are the outcome of such adaptations. Of prime interest are the metabolic adaptations which accompany the decreased thyroid hormone level and the effect of the latter on the genome. One example of such adaptation is the shift of the cardiac myosin isoenzyme from the fast form to a slow, highly efficient isoform. This transformation is compatible with the general decrease in metabolic rate. We may hypothesize that there may be similar changes in other ATPases.

5.2. Changes in signal transduction pathways

5.2.a. Cholinergic muscarinic pathway for water secretion
Alterations of the regulatory span involve modulations in signal transduction pathways for effector activation; they are poorly understood. Alterations in the signal transduction pathway for water secretion, the cholinergic-muscarinic transduction pathway, have been studied extensively in the rat submaxillary salivary gland. These include upregulation of the muscarinic receptors, altered handling of the neurotransmitter-evoked Ca^{2+} signal which is activated distal to the Ca^{2+} secretory processes, and augmented efflux of chloride, a marker for water secretion. There are some indications that there are conformational changes of the receptors or their G proteins. Possible benefits of these changes at different sites along the signal transduction pathway are modulation of the regulatory span and augmented secretion. These changes, coinciding with glandular hypertrophy,

collectively improve this organ's responsiveness and capacity. Hypertrophy of sweat glands and augmented water output per unit length of the secretory epithelium have been documented in patas monkeys. An increased glandular output-to-autonomic signal ratio is also evident in human sweat glands. These data imply that adaptation of the sweat glands resembles that of the salivary glands.

5.2.b. Adrenergic responsiveness and nitric oxide
In the cardiovascular system, studies on heat acclimation-induced membranal and cellular changes have been mostly confined to animal experimental models. In the vasculature, heat acclimation induces downregulation of the β-adrenergic response. This is accompanied by increased α-adrenergic receptor responsiveness and increased force generation by the smooth muscles of the blood vessels. β-Adrenergic stimulation induces vasorelaxation while α-adrenergic stimulation induces vasoconstriction. Thus, in the intact animal these changes in adrenergic responsiveness may be interpreted as an increased effector output-to-autonomic signal ratio. Concomitantly, at least in rats, a contribution of nitric oxide to increased splanchnic perfusion is also observed. The involvement of cytosolic Ca^{2+} in the adaptations observed in the cardiovascular system has been studied only in the heart, in conjunction with mechanisms of heat acclimation-induced increased pressure generation. The myocytes recruit a higher level of cytosolic Ca^{2+} for the contractile apparatus from the sarcoplasmic reticulum pool.

5.3. Cellular changes in central structures
A limited number of studies have shown an acclimatory response in neurons in the preoptic area of the hypothalamus: a reduced number of warm-sensitive and disappearance of cold-sensitive neurons have been observed. The significance of this change has not yet been evaluated.

5.4. Expanded optimal temperature margins via increased stock HSP level
An important aspect of membranal or cellular heat adaptation is associated with enhanced tolerance of higher core temperatures. In rats, both the cardiac muscle and the blood vessels display expanded optimal temperature margins. These findings match observations of improved resistance at higher core temperatures observed in the intact acclimated body. The responsible mechanism is not yet understood, but it is likely that changes occurring in protein properties and their thermal resistance contribute to this response. An important finding in this respect is that heat acclimation leads to an increased "stock" of the inducible heat shock proteins (HSP) from the HSP 70 kDa family. This provides protection to proteins by blunting protein unfolding upon stress ('chaperon' action). Other cytoprotective proteins such as antioxidants also play a role.

B-5

6. The Dynamics of Heat Acclimation: A Biphasic Model of Central-Peripheral Cross Talk

Day to day measurements of the responses to heat acclimation show that heat acclimation is a continuum of processes, varying temporally and differing in their efficiency and optimal performance. An apparent acclimated state is evident after a very short period of exposure to the new environment. It is brought about, however, by a cascade of transiently recruited mechanisms to alleviate the initial strain, even to the disregard of body homeostasis. Later, these initial mechanisms are replaced by long lasting processes, leading to an optimized and stable adaptive state. This is when acclimation has been achieved.

A classical example of these temporal variations is the overall acclimatory response of humans to humid heat. In humans, plasma volume expansion is an early acclimatory event. Preceding this event, cutaneous vasodilatation leads to increased heart rate and decreased stroke volume. Thus, plasma volume expansion allows an increase in stroke volume and, in turn, a decrease in heart rate. With progression of the acclimation process, cardiovascular stability, displayed by augmented stroke volume and lowered heart rate, is attained. At this stage, evaporation starts to increase, diminishing the need for highly elevated cardiac output and marked plasma expansion. The ultimate outcome is the classical acclimatory response described in *2.1*. Comprehensive studies on the dynamics of heat acclimation in animal experimental models, in which whole animals, and organ, cellular and molecular responses were studied in parallel yielded a biphasic heat acclimation model comprising short term and long-term phases of acclimation (Fig. 3).

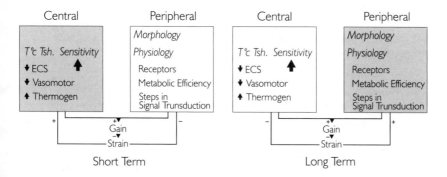

Figure 3. Heat acclimation - a conceptual model. During short-term heat acclimation, autonomic discharge overrides impaired organ responsiveness. During long-term heat-acclimation, the share of peripheral cellular adaptive features increases, thus accelerated autonomic discharge diminishes. Redrawn from [1].

The initial phase, short-term heat acclimation, constitutes transient processes. Impaired responsiveness of many target organs (e.g., evaporative cooling, atrium of the heart) due to desensitization of the receptors or other sites along the signal transduction pathway leads to accelerated autonomic excitation in order to maintain adequate effector output (Fig. 4). The autonomic nervous system thus plays the major role in alleviating the sustained strain.

In contrast, during long-term heat acclimation, the evoked, intrinsic peripheral adaptations reduce the need for accelerated autonomic excitation. The common denominator of the adaptive processes is increased efficiency, manifested by increased effector output despite decreased autonomic stimulation. It is likely that, during the evolution of this acclimation phase, cellular and molecular mechanisms play a major role.

Based on the data discussed in this section as well as in 5, it is evident that cellular responsiveness seems to play a modifying role in the central-peripheral cross talk, as demonstrated in Fig.4.

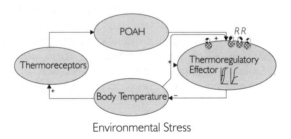

Figure 4. Central-peripheral interaction during heat acclimation. Temperature exerts its effect on both thermoreceptors and effector cell membranes; the latter effect modulates autonomic excitation. Redrawn from [1].

7. Loss of Acclimation and Its Re-Induction

Loss of acclimation and re-acclimation have been studied less than the process of heat acclimation. The operating mechanisms are, therefore, not yet determined. Complete loss of acclimation takes about three to four weeks. Although not linear, the process is gradual and not simultaneous for all effectors, nor for several features within the same effector. For example, in humans, the loss of acclimation is faster for heart rate and sweating rate than for rectal temperature. Thus, complete deacclimation for heart rate and sweating takes 3 weeks while, during this period, loss of acclimation for rectal temperature is only 50%. Physically fit individuals lose acclimation more slowly than unfit subjects.

The decay of acclimation has also been studied in the cardiac muscle of rats. Preacclimation pressure was restored after two weeks, while other features were lost after about three weeks.

In contrast to acclimation and deacclimation, re-acclimation is much faster and can be induced within 2 days. These observations are of practical importance for humans working in hot environments.

8. Problems Concerned with Human Acclimation to Heat

The interpretation of human acclimatory responses to heat faces several problems not usually encountered when animal experimental models are employed. This difference occurs because studies on human acclimation have practical implications, such as improved performance in hot environments. Therefore, most studies of human acclimation to heat have used combined heat acclimation and exercise training protocols. Each of these stressors, individually as well as in combination, elicits similar adaptive responses. For example, both stressors induce bradycardia and a larger blood volume. However, heat acclimation leads to improved cardiovascular reserve in the face of decreased metabolic rate, and exercise training results in improved cardiovascular reserve coinciding with increased metabolic rate. It is difficult, therefore, to interpret whether the improved endurance is attributable to both or to either one of the individual stressors. Furthermore, exercise intensity, the physical state of the subjects (trained *versus* non-trained, endurance *versus* short runners), the type of environment, e.g., hot and dry *versus* hot and humid, all affect acclimation. Acclimation to heat, therefore, may take a different course depending on the protocol used. This leads sometimes to controversies and conflicting conclusions. Although comparisons and interpretations are sometimes difficult, this plurality of protocols has contributed to our understanding of acclimation to multimodal stressors and to a variety of conditions which are beyond the scope of the present text.

9. Comparative Aspects of Heat Acclimation/Acclimatization

Despite the variety of species and taxonomic groups studied, the basic principle of heat acclimation is similar in nature for all species: acceleration of heat dissipation and reduction of heat production. The most intensively studied feature in all species is energy metabolism. This feature, which provides a general measure of metabolic status, shows a uniform pattern of adaptation. Adaptations in the cardiovascular system are also uniform. All species studied show decreased heart rate upon acclimation. Unfortunately, studies on peripheral blood flow have been carried out in only a limited number of species and almost no comparative data are available. Regarding the evaporative cooling system of homeotherms, there are a variety of mechanisms, each showing a somewhat different pattern of acclimation, particularly at the integrative level. The differences between salivation and sweating have already been discussed in *2.2*. Another mode of evaporative cooling is the panting mechanism. Acclimatory response of this mechanism in mammalian species has not yet been investigated. It has been studied, however, in birds. It was investigated in the rock pigeon in conjunction with its relationship to cutaneous evaporation. Rock pigeons, as well as several other bird species, use cutaneous evaporation as a regular component of their cooling repertoire, together with panting and gular fluttering. Upon heat acclimation, cutaneous evaporation becomes preponderant and, upon heat stress, the pigeon employs this mechanism instead of panting and gular fluttering. The benefits gained are decreased heat production by the respiratory muscles and better regulation of acid-base balance, which is highly impaired by increased respiratory rate. It is noteworthy that cutaneous evaporation is a neurally controlled mechanism and involves transient endothelial gap formation and accelerated plasma protein extravasation in the evaporating areas.

Acclimatory responses in homeotherms are all directed toward maintaining homeostasis. Acclimation is acquired at almost unchanged body temperature. In contrast, poikilotherms rely on the ambient temperature. Upon seasonal changes, cells and tissues of those poikilotherms which do not inhabit tropical regions are subjected to a wide range of temperature changes. The modes of physiological, biochemical and molecular acclimation of these species have attracted many investigators and our knowledge of these processes is wide. Compensatory responses include adjustment of the metabolic rate to the acclimation temperature (lower at higher ambient temperatures), transformation of a variety of isoenzymes to more energetically efficient isoforms, sometimes with altered affinity for the substrate, changes in synaptic transmission, and adjustment of membrane fluidity to the surrounding temperature (homeoviscous adaptation). It is noteworthy that acclimatory responses such as redistribution of myosin isoenzymes from fast to slow myosin, reduced contraction velocity, and adjustment of membrane fluidity are observed in both lower vertebrates (e.g., carp) and homeotherms (e.g., rat). It is, therefore, likely that at least some of the cellular/biochemical acclimatory responses to heat are common to all species.

10. Summary

10.1. Heat acclimation is attained by continuous or repeated exposures to the environmental stress. It is a highly complex process, with physiological adjustments occurring at all levels of body organization, ultimately leading to increased heat endurance and improved temperature tolerance.

10.2. In terms of the thermoregulatory effectors, global acclimatory response is achieved by decreased heat production, lowered heart rate, elevated stroke volume, augmented thermoregulatory skin blood flow, and increased capacity of the evaporative cooling system. These are brought about by concerted central and peripheral changes.

10.3. The process of heat acclimation is a continuum of processes, varying temporally and differing in their efficiency and optimal performance. It can be delineated as a biphasic model. The initial phase is manifested by a cascade of transiently recruited mechanisms to alleviate the initial strain, with the autonomic nervous system playing the major regulatory role. When acclimation has been achieved, long-lasting processes, including molecular, cellular and integrative adaptations that lead to an optimized and stable adaptive state, play the major role.

9.11. Suggested Readings

1. Horowitz, M. (1994) Heat stress and heat acclimation: the cellular response-modifier of autonomic control. In *Integrative and cellular aspects of autonomic functions: temperature and osmoregulation.* Pleschka K and Gerstberger R eds, John Libby eurotext, Paris, pp 87-95.

2. Pandolf KB, Burse RL, Goldman RF (1977) Role of physical fitness in heat acclimatization decay and reinduction. *Ergonomics,* **20**:399-408.

3. Senay LC, Mitchell D, Wyndham CH (1976) Acclimatization in a hot, humid environment: body fluid adjustments. *J. Appl. Physiol.,* **40**: 786-796.

4. Somero GN (1995) Proteins and temperature. *Annu. Rev. Physiol.* **57**: 43-68.

5. Horowitz M, Samueloff S (1989) Dehydration stress and heat acclimation. In *Milestones in environmental physiology.* Yousef M ed. *Progress in Biometeorology,* **7**:87-95.

Self -evaluation questions.

1. What is the global acclimatory response?
2. What characterizes acclimation of the evaporative cooling system?
3. Which are the major features characterizing acclimation of the cardiovascular system?
4. How does body fluid compartmentalization adapt to prolonged heat?
5. Specify cellular adaptive responses.
6. Specify and characterize phases in heat acclimation, deacclimation and re-induction of heat acclimation.

Chapter 11.2

Learning Objectives
1. Know and differentiate among the several adaptive mechanisms that allow endotherms to survive in the cold.
2. Define the terms that characterize the patterns of short- and long-term responses of endotherms to cold.
3. Describe the various strategies available to endotherms for survival in cold climates.
4. Know the physiological mechanisms that operate to conserve body heat during acute and prolonged exposure to cold.
5. Know the physiological mechanisms that operate to enhance body heat production during acute and prolonged exposure to cold.
6. Describe the various behavioral and autonomic adjustments that allow endotherms to save energy in cold environments.
7. Understand the hypothalamic modifications that underlie cold tolerance adaptation.

Bullets
B-1 Survival in very cold environments is facilitated by various adaptive strategies.
B-2 The "range of regulation" depends on several factors.
B-3 Different adaptive mechanisms exist to improve survival in the cold.
B-4 The "bioclimatic laws".
B-5 Cold-adaptive changes occur above and below the skin.
B-6 Shivering is the primary mechanism of heat production in the cold.
B-7 The principal source of NST is BAT.
B-8 Shivering and nonshivering thermogenesis are interdependent.
B-9 Certain animals allow their temperature to fall better to survive in the cold.
B-10 Different adaptive reactions underlie heterothermy.
B-11 Widening of the interthreshold zone is common to most forms of dormancy.
B-12 Thermal and modulatory inputs are integrated in the hypothalamus.
B-13 Various neuroactive substances released in the hypothalamus provide modulatory inputs to the thermoregulatory system.
B-14 Hypothalamic norepinephrine and serotonin importantly influence the central integration of thermal signals.
B-15 Behavioral adjustments contribute in a major way to survival in the cold.

Chapter 11.2

COLD ADAPTATION

EUGEN ZEISBERGER
Physiologisches Institut, Klinikum der Justus Liebig Universität
Aulweg 129, D-35392 Giessen, Germany

1. Introduction

This chapter will describe the physiological mechanisms that endotherms (mammals and birds) use to survive decreases in the temperature of their environment. In the face of moderate cooling, these animals generally maintain their body temperature relatively constant at different temperatures between 35°C and 40°C (depending on species) by reducing their heat loss and increasing their heat production. Eventually, however, the capacity for heat production may become insufficient to counteract the cold load, and the temperature of the body core decreases to an extent incompatible with normal functioning of organ systems such that the animals may die. The range of environmental temperatures in which an animal can survive is called its range of tolerance. At the extremes of the tolerance zone are ranges of resistance, in which the normal organ functions are disturbed, and the animal will eventually die. The animals reach this range rarely, only in cases of very sudden and extreme changes of environmental temperature. Prolonged or intermittent decreases of ambient temperature induce improvements of the capacity for heat production and body insulation, which shift the resistance range so that the animal can survive lower temperatures of the environment. Such changes, which prepare the animal to better resist the cold, are called cold adaptation. Depending on what factors induce the process of cold adaptation, it is differentiated between acclimation (modifications induced in the laboratory solely by temperature changes) and acclimatization (modifications induced in nature by several factors like the seasonal patterns of daily fluctuations in temperature, air velocity, humidity and the length of the day).

It should be stressed that the adaptive changes are complex, and include behavioral, functional and morphological modifications that developed slowly during the process of evolution. Different species use different strategies of adaptation to cold, which they have inherited from foregoing generations. It can be said that the adaptive process involves learning, and may be modified by individual experience. These complex specific behaviors are governed by autonomic centers in the hypothalamus and limbic system of the brain. Here, the signals from peripheral and central thermosensors, referring temperature changes, are integrated with signals of other homeostatic hypothalamic systems to generate emotions and behavioral drives. In addition to appropriate vegetative reactions reducing the acute cold stress, the stream of consciousness may be dominated by such drives and the actions are oriented to a physiologically apt behavior.

1.1. What temperatures are cold?

Measurements of static activity of peripheral cold sensors and central warm sensors demonstrated that both are continuously active in a thermoneutral environment, although no conscious temperature sensation is felt. Thus, the temperature sensation does not reflect the activity of peripheral or central thermosensors alone, but rather a state derived after integration of these signals in the thermoregulatory system. The thermal sensation should be differentiated from emotional feelings of thermal comfort or discomfort, which are either "pleasant" or "unpleasant" and are the result of a more complex integration (See also Chap. 6 and 7).

Theoretically, all temperatures below the lower limit of the thermoneutral zone (TNZ) can be denoted as cold, because they initiate regulatory cold defence mechanisms such as an increase in metabolic rate. Since some cold defence mechanisms, like piloerection, peripheral vasoconstriction, and behavioral changes in body shape, are activated earlier, it can be concluded that already ambient temperatures below the upper limit of TNZ should be regarded as cold. It is known that the width of TNZ differs in species according to their insulation [5]; i.e., cold sensation is related to heat loss. Since this varies depending on body size and insulation, it follows that cold sensations are induced at different ambient temperatures in different species. They change also within the same species during growth. As a rule, both the TNZ and the range of regulation (range of ambient temperatures within which the core temperature can be maintained constant by means of metabolic heat production) are narrower in the newborn (due to their large surface-volume ratio and a poor insulation) than in adults. For humans, this is illustrated in Fig. 1. (See also Chap. 9.1).

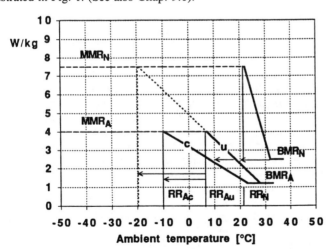

Fig. 1: Comparison of heat production as a function of the ambient temperature in a newborn(N), and an unclothed (u), and clothed (c) adult(A) human. Adapted from [6]. See text for explanation.

The basal metabolic rate (BMR) of an unclothed (u) adult human (1.2 W/kg) can balance the heat loss until the ambient temperature of 28 °C is reached (the lower limit of the TNZ), from where the heat loss prevails and must be compensated for by an increase in thermoregulatory heat production (by shivering). The maximal metabolic rate (MMR) for adult humans amounts to 4 W/kg. This metabolic level is thus achieved at an ambient temperature of 7 °C (the upper level of resistance range). In a clothed (c) adult human, the lower limit of TNZ is postponed to an ambient temperature of 22 °C (room temperature) and the upper resistance range limit (RR_{Ac}) to -10 °C. This shift is due to a change in insulation only, and not to a change in metabolism. In a human neonate, the BMR_N amounts to 2.5 W/kg, which can balance heat loss until the ambient temperature reaches 32 °C (down from 34 °C). Although the maximum metabolic rate in the neonate (MMR_N) is absolutely almost twice as high than in the adult, it is reached already at an ambient temperature of 22 °C (RR_N). Thus, in a neonate, both the TNZ (32°C - 34 °C) and the regulatory range (22 °C - 32 °C) are narrower than in an unclothed adult (28 °C - 32 °C and 7 °C - 28 °C, respectively). In a clothed human, the insulative value is improved by about 70 %, which enlarges both the TNZ and the regulatory range to the values quoted above.

1.2. Strategies for survival in the cold

In order to survive lower environmental temperatures, it is necessary to enlarge the regulatory range. The possible ways of how to achieve this in a human have been shown in Fig. 1. The upper range of resistance in an unclothed adult (RR_{Au}) can be shifted from 7 °C to -10 °C by reducing the heat loss due to the improvement of insulation by means of clothing (RR_{Ac}). Another possibility would be to increase the metabolic heat production. If the adult had the same MMR as the neonate, it would be possible to shift the RR_{Au} even to -20 °C (indicated by dotted lines). Unfortunately, the adult human does not have this possibility. The increased MMR_N due to nonshivering thermogenesis (NST) in brown adipose tissue (BAT) is irreversibly lost in adults. If the adult had a BMR as high as the neonate, the lower limit of the TNZ would shift from 28 °C to an ambient temperature below 20 °C in an unclothed human, and from 22 °C to about 9 °C in a clothed human (as indicated by the arrows in Fig. 1). Although it is not possible for an adult human to double his BMR (which would correspond to the BMR of the neonate), in two ethnic groups, the Eskimos and the Alacaluf Indians, increased BMRs in the order of 20%-40% have been described, which may decrease the lower limit of TNZ by 3 °C - 5 °C [1]. The last possibility (not shown in the figure) would be to decrease body temperature.

Endotherms thus have three main strategies for survival in the cold: 1) they decrease heat loss (by changing body shape, body size, or by improving their insulation)
2) they increase heat production (by increasing the BMR, by improving the efficiency of shivering, or by additional thermogenic mechanisms like NST)
3) they decrease body temperature (by changes in the thermoregulatory system in order to tolerate large variations of temperature in the body core).

All of these strategies may be used by endotherms when they adapt to cold. Different species use different strategies. Some, mostly small mammals, use all 3 strategies in a time sequence from 3 to 1. In the following, these strategies will be explained in more detail and a few typical examples will be given.

1.2.1. Strategies for decreasing heat loss

a) Changes in Body Size.

Body size is an important determinant of thermal conductance. Newborn animals have high absolute thermal conductance due to their large surface:volume ratio. By growing, they reduce this ratio and thus decrease heat loss (cf. Fig. 1). For the same reason, larger species have a lower conductance than smaller species, and their metabolic heat production is relatively less affected by low ambient temperatures. For the individual adult mammal, however, increased body mass is not necessarily a useful response to cold stress, because it cannot go up to such an extent as to be effective. Body mass would have to be increased by 50 % to decrease the lower limit of TNZ by only 2 °C, and by 100 % to decrease it by 3 °C. However, endotherms that live in cold climates often have a higher body mass than comparable individuals from warmer climates. This is known as Bergmann's rule, one of the bioclimatic laws postulated in the 19th century and can be interpreted as a genetic adaptation to the environment. Moreover, animals in cold climates tend to have smaller surface areas than animals of the same species in hot climates. For example, polar animals tend to be larger, heavier and have relatively short limbs or appendages, whereas tropical animals are tall and slim with long and thin extremities. This is known as Allen's rule, another of the bioclimatic laws [1, 2]. A change in body size takes a long time to manifest and can thus be realized only as a seasonal adaptation. During winter, endotherms have a greater need for energy to compensate for heat loss which, unfortunately, is concurrent with a reduced availability of food at this time of year. It seems that reduced availability of food is a factor modifying the strategies for survival. Large animals, above several kg of body weight, seem to rely on seasonal improvements of fur insulation, which allow them to maintain normothermia at a given ambient temperature with much less energy than during summer (e.g. 50 % in red fox or the reindeer). Small mammals cannot improve insulation to such an extent and thus depend upon metabolic modifications. They may use hibernation, which reduces total energy requirements by 85 %. Alternatively, winter-active small mammals may reduce energy requirements by a combination of winter weight reduction, slight improvements of thermal conductance, and shallow daily torpor. In the Djungarian hamster, this strategy may save 63% of energy as compared to summer-acclimated hamsters.

b) Changes in Insulation.

Thermal insulation primarily depends upon the low heat capacity and the low thermal conductivity of the layer of air trapped in the fur. It can be improved by increasing the depth of the fur and by adding a dense underfur which reduces air convection within the fur. This possibility can be used only by large mammals, since they can grow fur with up to 10 cm of hair length. Small mammals, with less than 1 kg body weight, can grow a fur of at most 1 cm in depth, in which thermal insulation cannot be substantially improved. Some mammals and birds seasonally alter their fur coat and at the same time change coat color to white for cryptic coloration. This change may reduce the conductance by up to 40 %. This can explain the reduction in energy expenditure of about 50% in comparison to summer, as mentioned above. The thermal resistance of the pelage in mammals and the plumage in birds can be increased, within certain limits, by raising of hair and feathers due to involuntary tensing of smooth muscles (arrectores pilorum) connected to the hair follicles (piloerection). Some mammals, especially those living or diving in cold water, develop additional thermal insulation in the form of a subcutaneous fat layer, that can substantially enlarge the buffer zone between the environment and the core of the body. The same purpose is served by the enlargement of the thermolabile body shell area by circulatory adjustments.

B-5

c) Circulatory Adjustments.
Skin blood flow is primarily controlled by adrenergic vasoconstrictor nervous activity (See also Chap. 5). In the cold, noradrenaline is released from peripheral sympathetic nerve terminals and induces not only a vasoconstriction of skin vessels, but also a redistribution of blood flow in the extremities and the body core. More blood is circulated in the temperature-stable body core, and less blood is sent to the extremities. Additionally, due to the countercurrent blood flow in central arteries and veins, excessive extremity heat loss is prevented in the cold. In warm environment or during activity, the blood returns from the extremities to the body core via superficial veins. Thus, the countercurrent heat exchange in central arteries and veins is interrupted, and more heat can be dissipated from the extremities to the environment. Moreover, the speed of the blood flow through the extremities and the skin can be modified by reflex activity. Along with an abundance of arterioles, significant numbers of arteriovenous anastomoses (AVAs) are extant in acral sites. By means of opening the AVAs, the blood flow, and therefore heat flux, through the acral sites can be substantially accelerated. These special arrangements of blood flow in acral sites are important not only for heat loss, but also for preventing the freezing of these ineffectively-insulated appendages at subfreezing air temperatures.

d) Evaporative Heat Loss.
In the cold, heat loss mechanisms are inhibited. Therefore, in humans, there is no sweating and the evaporative heat loss is restricted to insensible evaporation of water from the skin and the respiratory tract. Theoretically, the respiratory evaporation should increase at lower ambient temperatures, because the increased heat loss must be compensated for by a higher metabolic rate and, therefore, by increased ventilation. However, this tendency is minimized by nasal countercurrent heat exchange. When breezing cold air, the surfaces of nasal airways are cooler than the lung tissue, and vapor in the expired air can condense on them. In humans, this mechanism is not very effective. In smaller arctic animals that increase their respiratory rate considerably in cold environments, it can be so effective that they reduce their respiratory heat loss to below 5% of the total heat loss.

1.2.2. Strategies for increasing heat production

Heat production in endotherms is clearly related to body size. From a comparative point of view, the smallest mammal, the Etruscan shrew (5 g), produces about 175 times more heat per unit of body weight and unit of time than an elephant (about 4,000,000 g) to compensate for its heat loss at TNZ and to keep its body temperature normal. Thermoregulatory thermogenesis is stimulated when an animal is exposed to temperatures below its TNZ. The cold-induced increase in thermogenesis has three components:
1) long-term increase in basal metabolism
2) shivering thermogenesis in skeletal muscles, controlled by the cholinergic somatomotor nervous system, and
3) nonshivering thermogenesis (NST) in brown adipose tissue (BAT), activated by nerve fibres of the peripheral sympathetic system.

a) Long-Term Increase in Basal Metabolism.
Increases in basal metabolism by up to 40% as adaptations to cold have been demonstrated in mammals and birds. They are mediated by increased sensitivity to several hormones released during long-term exposure to cold. Well documented among these are increases in levels of thyroid and adrenal hormones that determine the rate of cellular metabolism and stimulate the development of NST.

b) Shivering.
Shivering is a myotactic reflex oscillation due to muscle spindle activation by efferents from γ-motoneurons (See Chap. 4.1). Shivering is controlled by tracts of the "shivering pathway" that originate in the posterior hypothalamus and descend to the lower brain stem to contact the supraspinal motor pathways. Shivering is the primary thermogenic response to cold in all endotherms. It is the exclusive source of cold-induced thermogenesis in birds and large mammals (above 10 kg body weight) at rest. The skeletal muscle has a high aerobic metabolic capacity (about 60 W/kg during maximal exercise) and represents a considerable proportion of the body mass. The maximal metabolic rate during short-time maximal exercise amounts to nearly 10 times of BMR in all species (cf. Fig. 2). In every-day life, exercise can substitute for NST in the cold. It can, of course, substitute also for shivering, but it is less economical, because it would increase thermal conductance. In resting animals, shivering must be kept up over longer time periods and, therefore, it increases the metabolic rate less than a short-time exercise (MMR). In humans, it is limited to 3 times BMR; in small animals, to 5 times BMR.

c) Nonshivering Thermogenesis.
NST is a second mechanism for augmented metabolic heat production. Most of it is generated in BAT which has been identified only in small placental mammals. BAT is present in some newborn mammals, some hibernating mammals, and some cold adapted mammals [cf. 3, 4 for reviews, see also Chap. 4.2]. The maximum amount of NST has been shown to be inversely related to body mass (Fig. 2).

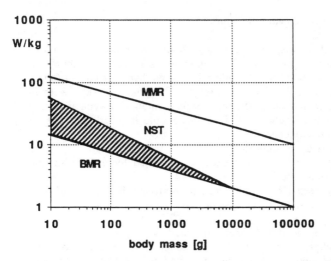

Fig. 2: The amount of nonshivering thermogenesis (shaded) in a number of rodents and other mammals with body weights ranging from 10 g to 10 kg (Cf. text for explanation). Adapted from [7],

The lower line indicates the predicted BMR, and the upper line the calculated MMR for short-time maximal exercise. The shaded area denotes the maximum amount of NST, measured as an increase in oxygen uptake following noradrenaline injection in a number of rodents and other mammals with body weights ranging from 10 to 10000 g [7].

BAT has a remarkable aerobic capacity, able to produce heat at a rate equivalent to 500 W/kg. This is an order of magnitude greater than the aerobic power of the muscle. The biochemical mechanism of conversion of substrate energy into heat is based on uncoupling of oxidation from phosphorylation. In order to meet the high oxygen requirement for thermogenesis, BAT has a rich vasculature, estimated to be 4-6 times as dense as that of white fat. This can support an up to 5-fold increase of blood flow through the tissue in cold environment, amounting to 25 % of the total cardiac O_2 delivery. In most mammals, NST is activated prior to shivering when they are exposed to cold. Vascular connections, which deliver warm blood from interscapular BAT to the spinal cord, are responsible for this postponement of shivering to the more extreme ambient temperatures. Thus, shivering remains inhibited so long as the interscapular BAT produces enough heat to keep the activity of the spinal warm sensors sufficiently high to inhibit the stimulatory input from peripheral cold sensors.

BAT is particularly adaptable. Its thermogenic capacity can be increased in two ways:

a) by acquiring more brown fat cells, i.e., growth of BAT (hypertrophy, which may or may not involve hyperplasia), and
b) by improving cellular thermogenesis (making "better" cells, with a higher number of mitochondria with more thermogenin).
Three stimuli have been shown to induce the growth of BAT: cold, short photoperiod, and eating of a highly preferred diet ("cafeteria" diet). A stimulated growth of BAT need not necessarily be accompanied by a markedly increased thermogenesis. It may be that the diet-induced thermogenesis and short photoperiod preadapt mammals for the use of NST in the cold. Although different classical growth factors support the growth of BAT, it seems that the main stimulator, of both the cell division and the gene expression of thermogenin, is noradrenaline released from sympathetic nerve endings. However, different receptors are involved. Whereas β_1-receptors are involved in BAT growth, β_3-receptors stimulate the acute thermogenic response and the thermogenin gene expression. Optimal thermogenic function of BAT also seems to require the intracellular conversion of thyroxin to triiodothyronine [cf. 3, 4 for reviews].

1.2.3. *Strategies for modifying the thermoregulatory system*

1.2.3.1. *Heterothermy*

Heterothermy is a generally used strategy enabling the organism to save energy. B-9

a) Peripheral or Regional Heterothermy.
Endotherms with large body size allow the temperature of peripheral tissues, such as skin, extremities and appendages, to decline below that of the body core. This is due to the circulatory changes described above. During adaptation to cold, these changes may be more profound because the sensitivity to neural and humoral mediators may be increased.

b) Temporal Heterothermy.
Many endotherms allow that the temperature of their body core declines as well, and the animals become temporarily hypothermic. There is a confusing terminology for various types of hypothermia. Hypothermia may be natural or experimentally induced. The common criterion for different forms of natural hypothermia is that the animal is able to spontaneously regain (arouse, rewarm) the normal temperature of its body core, using endogenous heat production. Depending on the depth of the hypothermia, the following forms of temporal heterothermy can be differentiated:
a) adaptive hypothermia - mild natural hypothermia,
b) torpor - pronounced natural hypothermia,
c) hibernation - long-term torpor in response to winter cold and food deprivation.
All these forms of temporal heterothermy involve changes in regulatory characteristics, which can be defined as deviations in the threshold and gain of the thermoregulatory responses. These can be further subdivided into adaptive and genetically fixed changes (torpor and hibernation). B-10

c) Adaptive Hypothermy.

A number of birds and mammals use moderate hypothermia (core temperature remains above 30 °C) as a short-term response to cold stress. It can be subdivided into short-term (minutes) tolerance adaptation and long-term (weeks) tolerance adaptation [cf. 3, 4 for reviews].

For example, short exposure to a cold environment evokes not only thermoregulatory reflexes and emergency stress reactions, but results in a temporary shift of the shivering threshold to a lower mean body temperature (by several tenth of 1 °C) which persists for about 30 min after rewarming. This is called a short-term deviation of the thermoregulatory response. Continuous confrontation with a cold environment for several weeks results in a more pronounced shift of the threshold for heat production to a lower body temperature (up to several °C), which can be called a long-term modification of the thermoregulatory characteristics. Alternatively, when the threshold temperature for onset of heat loss does not shift so much, a widened interthreshold zone (ITZ) often results, i.e., the animal tolerates larger fluctuations of body temperature. The precision of the control system is lowered, but this change is highly economical because it saves much energy. Some animals living in extreme climatic conditions are known to have a wide interthreshold zone (in camels from 3°C to 6°C, which is a very useful adaptation to desert life because it saves not only energy but also water).

d) Torpor

Torpor is a pronounced natural hypothermia accompanied by a substantial depression of metabolism, respiratory rate, heart rate, and lack of motor coordination and of response to external stimulation. Similar changes, but to a much lesser extent, can also be seen during normal sleep. Torpor is used mostly by small endotherms during periods of inactivity. Due to their high rates of metabolism, they would be subjected to starvation during periods of inactivity, when they are not feeding. Most species enter into torpor during the night; some bats even enter into periods of short-time torpor during the day when inactive. Before the animal becomes active, its body temperature rises as a result of a burst of metabolic activity, mainly induced by oxidation of BAT stores. Torpid animals must hide, because they cannot escape if detected by predators. Therefore, some animals use torpor only when energy reserves are very low.

e) Hibernation

Hibernation is used to describe extended periods of torpor. Although many small endotherms undergo a daily cycle of torpor, only few are true hibernators. Their stored energy reserves would be quickly consumed even in the hibernating state. All true hibernators are mid-sized mammals (Rodentia, Insectivora and Chiroptera), having stored sufficient substrate reserves for extended hibernation. But there are no true hibernators among large mammals. Bears were thought to hibernate, but they simply enter into a winter dormancy, with only a moderate decline in body core temperature (31 °C - 35 °C) and in metabolic rate. Hibernation seems a perfect strategy to cope with a shortage of energy supply during winter. The true hibernators retreat into a frost protected burrow or cave for up to 7 months. They lower their body temperature close to the ambient temperature, thus reducing energy requirements to only 5 - 15% of what is needed during activity. Most of the energy required during hibernation is spent for the repeated arousals which occur at about weekly intervals. Hibernation must be viewed as an economically highly specialized behavior.

Animals cannot hibernate any time. The readiness for hibernation depends on annual and daily rhythms. It is restricted to the winter season, and the animals enter into hibernation during the time of the day in which they normally sleep. Hibernation seems to be inhibited by gonadal activity. The animals do not enter hibernation as long as the gonads are active, and finish hibernation with the start of gonadal activity in the spring. Animals are slowly prepared for hibernation by a set of internally elicited actions, including preparing winter quarters and hoarding food. The onset of hibernation is often preceded by increasing circadian fluctuations of body temperature ("test drops"). The final entry into hibernation is a slow process of "controlled cooling", taking 8 to 18 hour depending on species. Arousal from hibernation requires less than 3 hours. The process of arousal is an extremely dramatic physiological event. It is characterized by rapid increases in heart and respiratory rates and oxygen consumption. The rapid warming is due to intensive oxidation of brown fat in the first phase of arousal, which is then accompanied by shivering leading to an overshoot in metabolic rate in the second phase of arousal. During torpor and hibernation, thermoregulatory control is not suspended, but cold defence mechanisms continue with lowered threshold temperatures. During hibernation, the threshold temperature for the onset of NST may be reset to more than 25 °C below normal. At ambient temperatures above 5 °C, the hibernators keep their temperature as low as 1 °C above ambient. If the air temperature falls to dangerously low levels, the animal increases its metabolic rate to maintain a constant, low body temperature or becomes aroused. The functioning of the thermoregulatory system in such low temperatures is possible due to some specializations of hibernators. Apparently, the thermal sensitivity of the skin and the function of at least some parts of the central nervous system are maintained during hibernation, and respiration as well as circulation are slowed, but functioning at low temperatures (i.e., the animals avoid somehow the heart fibrillations seen in nonhibernating species at temperatures between 19 °C and 25 °C). The parts of the central nervous system coordinating vigilance and the activity of motor systems are switched off or inhibited.

All forms of dormancy like ordinary sleep, torpor, hibernation, "winter sleep", and estivation are neurophysiologically and metabolically similar. The changes in the regulatory system can be described as an extreme widening of the interthreshold zone (ITZ). The mechanisms leading to the extreme widening of ITZ in hibernators are not known. It was speculated that during hibernation thermoregulation is no longer controlled by hypothalamic, but rather by lower brain stem centers that, again, switch on the hypothalamic centers during arousal [1]. The next paragraph will explain how the width of the ITZ can be changed at the hypothalamic level of control during tolerance adaptation.

1.2.3.2. Modulation of thermoregulatory thresholds during
tolerance adaptation

a) **Organization of the Thermoregulatory System.**
The properties of the thermoregulatory system can be summarized in a most simple way as follows (See also Chap. 6). The thermoregulatory system is organized hierarchically, the highest controller being localized in the hypothalamus. The central controller receives information on the temperature from the body surface and from different parts of the body core. Body surface temperature is measured by cutaneous thermal sensors, core temperature by thermosensitive structures within the preoptic area in the limbic septum, anterior hypothalamus, brain stem, spinal cord and possibly some other places. Although two types of sensors, warm and cold sensitive structures, have been found in the skin and in the body core, it seems that in afferent nerve fibres from the skin the information from cold sensitive structures dominates, and on the other hand, among the thermosensitive structures of the body core, only few are cold sensitive. In cybernetic terms, two signal inputs into the hypothalamic controller are considered:
1. The composite skin temperature signal (from cutaneous cold receptors), combined with inputs from the few central cold sensitive structures, and
2. The composite core temperature signal (from the large majority of warm-sensitive central structures), combined with inputs from cutaneous warm sensors.

These inputs are coupled to effectors in such a way that the composite skin temperature signal from cold sensors activates the mechanisms for heat conservation and production, and the composite core temperature signal from warm sensors activates the heat loss mechanisms. As known from various experimental approaches, the cold input may inhibit the heat defence branch and vice versa. Therefore, inhibitory neurones have been proposed in the central controller, interconnecting the two parallel branches for cold and heat defences. In this way, activation of cold defence can be evoked only after overcoming the inhibition from central warm sensors, and vice versa, heat defence can begin after the input from cold sensors is surpassed. Thus, the activation of thermoregulatory responses starts at certain combinations of central and peripheral temperatures = threshold temperatures [cf. 3, 4 for reviews]. In an attempt to imitate the hypothalamic integration of thermal signals, some authors relate thermoregulatory responses to mean body temperature (Tb), which is a species-specific combination of a representative central temperature and the mean peripheral temperature. The weighting factors determining Tb seem to vary with body size, ranging from 0.4:1 to 1:1 (central : mean peripheral temperature) in small animals like rats or guinea pigs, 3:1 in rabbits, 4:1 in dogs, and from 8:2 to 9:1 in humans [8]. This relates to the relatively large surface area in a small animal as representing an important thermoregulatory input, the importance of which diminishes in large organisms. A simple attempt to illustrate the integration of signals in the hypothalamic controller is shown in Fig. 3.

B-12

The shaded arrow in the left inset represents the thermal characteristics of the composite core temperature signal. The dark arrow in the right inset represents the thermal characteristics of the composite skin temperature signal. The central inset illustrates the hypothalamic integration of these signals. After surpassing the inhibitory action of the opposite input, increasing signals (represented by shaded areas) can be created by changes in the mean body temperature, activating either cold defence or heat defence mechanisms. Such an arrangement would result in a stable thermoregulatory set-point. Since it has been known that between the thermoregulatory threshold temperatures for the onset of heat production and dissipation a zone rather than a point exists, additional modulatory inputs into the controller are inferred.

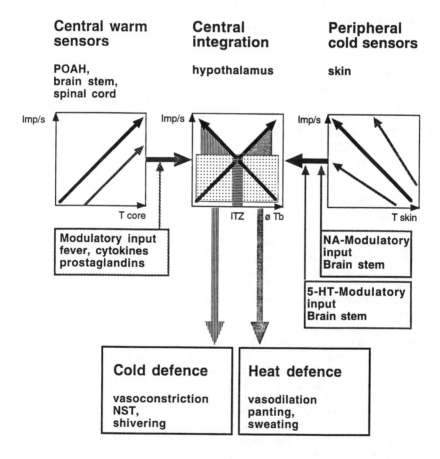

Fig. 3: A simple model illustrating the way of integration of thermal signals with modulatory inputs in the hypothalamic controller. For explanation cf. text.

b) **Modulatory Inputs into the Thermoregulatory System.**
Three main modulatory inputs into the controller are shown in the model in Fig. 3:
1) The thermal signals are compared with non-thermal signals determining the basal activity state (represented by the dotted area in the central inset in Fig. 3), which must be exceeded by thermal signals in order to create output signals driving the thermoregulatory effectors. This input is comparable to a discriminator window hiding the signal noise and allowing the establishment of a set range (ITZ) instead of a set point.
2) The composite skin temperature signal can be modified at the hypothalamic level of integration by opposite modulatory inputs from noradrenergic (NA) and serotonergic (5-HT) neurons in the lower brain stem. These additional inputs shift the composite skin temperature signal to the right or left, as indicated by the shaded arrows in the right inset of Fig. 3. Such modifications lead to corresponding shifts in thresholds for cold defence mechanisms and to changes in slopes of thermogenic responses to cooling.
3) The composite core temperature signal (shown in the left inset of Fig.3) can be shifted to the right (as indicated by the dark shaded arrow) by modulatory substances (e.g., cytokines and prostaglandins) released in the hypothalamus during fever and some other physiological situations. Such a modification leads to a corresponding shift in the threshold for heat defence mechanisms.
All these modulatory inputs can be activated to a different extent in different physiological situations, resulting in modifications of the thermoregulatory system, characterized by widening or narrowing of the ITZ, with corresponding changes in the slopes of thermoregulatory responses to external cooling or warming, and leading eventually to a change in body temperature, sometimes interpreted as a changed "set-point" [cf. 3, 4 for reviews].

c) **Experimental Modifications of Thermoregulatory Thresholds.**
In many species, changes in regulatory characteristics have been described during ontogenetic development and adaptation to warm or cold climates. Much experimental evidence supports the participation of antagonistic brain stem modulatory inputs in deviations in threshold and gain of thermoregulatory responses to experimental cooling or warming.
Figure 4 shows an example of such modifications, based on our own experiments in guinea pigs. This figure summarizes the reevaluation of previous experiments in which the responses to experimental cooling or warming were tested after intrahypothalamic (i.h.) microinjections of NA or 5-HT [9], or after neurochemical lesions of these afferents by injections of specific neurotoxins into the areas in the lower brain stem from which these pathways originate [10]. The horizontal lines connecting the metabolic rate of 5 W/kg and respiratory evaporative heat loss (REHL) of 1 W/kg denote the normal values measured at thermoneutral ambient temperatures. The different lines denote the responses to changes in mean body temperature evoked by experimental cooling or warming, indicating the threshold temperatures for the onset of cold defence (increase in MR) or heat defence (increase in REHL) and the slopes of these reactions. The shaded bars indicate the width of the ITZ.

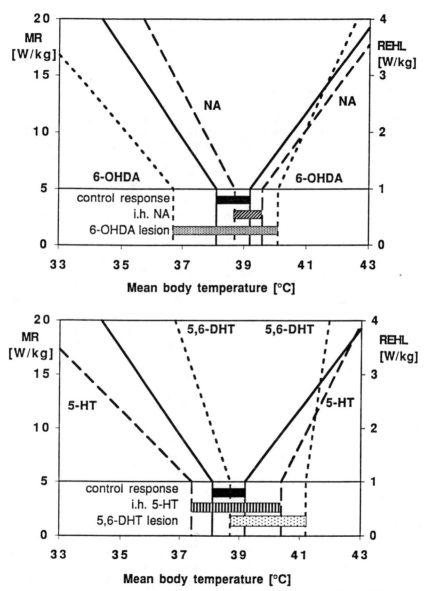

Fig. 4: Effects of intrahypothalamic injections of noradrenaline (NA) or 5-hydroxytryptamine (5-HT) or of neurochemical lesions of these afferents by injections of neurotoxins, 6-hydroxydopamine (6-OHDA) or 5,6-dihydroxytryptamine (5,6-DHT), into the lower brain stem on responses of guinea pigs to experimental cooling or warming. Adapted from [9, 10]. See text for explanation.

The upper part of Fig. 4 summarizes the manipulations of the noradrenergic input. The lower part indicates the effects of manipulations of the serotonergic input. This comparison demonstrates opposite effects of NA and 5-HT (denoted by dashed lines) on the cold defence reactions. Both effects of i.h. microinjections of NA and 5-HT on cold defence reactions could be reversed by selective lesions of the respective inputs by neurotoxins 6-hydroxydopamine (6-OHDA) and 5,6 dihydroxytryptamine (5,6-DHT) as denoted by the dotted lines. Therefore, it was assumed that these inputs modulate the composite skin temperature signal in the model shown in Fig. 3.

The injections and lesions influenced also the heat defence reactions. However, both NA and 5-HT microinjections increased the threshold temperature for REHL, and these changes could not be reversed by selective lesions of these inputs. Therefore, it seems that the increased threshold temperatures for heat defence reactions are due to hypothalamic release of prostaglandins, which have been proposed to modulate the composite core temperature signal in Fig. 3. Prostaglandins may be released by mechanical stimuli induced by intrahypothalamic microinjections [11]. Their production may also be controlled by the activity in NA and 5-HT inputs; thus, it may be disinhibited by selective lesions of these inputs, resulting in an increased threshold temperature for the onset of heat defence reactions and in a higher sensitivity to the increase in body temperature. The hypothalamic release of prostaglandins may also be stimulated by cytokines, the signal substances of activated immune cells mediating the febrile modification of the thermoregulatory system.

Thus, the changes in the thermoregulatory system described during ontogeny, thermal adaptation, and fever can be explained, in the guinea pig, by the existence of these modulatory inputs to the central integration of the thermal signals. The width of the ITZ depends of the balance between the modulatory NA and 5-HT inputs. A narrow ITZ with an average of body temperature at 39 °C results when the NA input dominates (situation found in warm-adapted guinea pigs). A wide ITZ with an average of body temperature at 38 °C results when the 5-HT input dominates (situation found in newborn and cold-adapted guinea pigs) [cf. 3, 4 for reviews]. The body temperature may temporarily increase to 41 °C during fever due to a modulation of the thermoregulatory system by cytokines and prostaglandins. All these changes in guinea pigs are relatively small and may modify the mean body temperature within a range of, at most, 5 °C.

In humans, the ITZ extends from 36.0 °C to 36.6 °C, on average. Its width can be nearly doubled in cold-adapted humans, mainly due to a shift of threshold for shivering to a mean body temperature of 35.4 °C. This cold-induced modifications may be due to changes in the activity of modulatory inputs from the lower brain stem, described under b). General anesthesia can widen the ITZ much more, extending it from 33.0 °C to 38.0 °C [12]. The reasons for the widening of ITZ during anesthesia have not been explained as yet. It seems that anesthesia resembles a situation in which the non-thermal signals inhibiting the interactions of thermal signals (dotted area in the central inset in Fig. 3) were augmented. The third modulatory input, also described under b), together with an activation in the NA-brainstem afferents, is responsible for a shift of ITZ to higher mean body temperature during fever. The increase in temperature of the body core may be as high as 41 °C during high fevers.

1.2.3.4. Components of the cold adaptive process

The whole cold adaptive process consists of several components that have different time requirements and overlap each other. In the guinea pig, for example, they can be described as follows.

Acute cold exposure activates the sympathetic system, thereby actuating immediate control actions like piloerection, vasoconstriction, and mobilization of metabolic substrates for muscular contractions. The first response to intensive cooling is exaggerated and uneconomical. Cold is a stressor inducing a nonspecific stress response, or emergency reaction, leading to an activation of the sympatico-adrenal system and the hypothalamic-pituitary-adrenal hormonal axis. The high release of hormones and a continuing increase in peripheral NA levels, due to an overspill from activated sympathetic nerve endings, constitute feedback signals leading to a change of the thermoregulatory system [cf. 3, 4 for reviews], described as an adaptive hypothermia. The threshold temperature for shivering (the only thermogenic mechanism in a warm-adapted guinea pig) is shifted to a 1.2 °C lower mean body temperature than in normal animals. The widening of the ITZ allows the animal to tolerate larger deviations in body temperature and thus save energy. In the long run, the increased peripheral NA levels, together with other hormones, promote the development of BAT and NST. This process takes about three weeks. As the threshold temperature for NST is higher than that for shivering, there is a tendency to reestablish normothermic temperature control. In prolonged exposure to cold, body insulation is improved by growth of fur. Consequently, the external cold will no longer be stressful. This may lead to an involution of the preceding adaptive components.

1.2.3.5. Behavior

Behavioral responses are energetically very economical, and animals use them preferentially, both in cold and hot environments [cf. 4 for reviews]. Hence, in addition to autonomic responses, exposure to cold sets in motion behavioral reactions (See also Chap. 7). These can be very different in different species and include the following groups of responses:

a) Postural Changes

A sphere has a minimum surface/volume ratio, hence it is the energetically most effective shape. For example, mammals may curl up and birds may retract their head and fluff up their feathers to assume a spheric shape. To minimize convective heat loss, some animals huddle or aggregate with other members of the group. In deer mice and voles, huddling can reduce their energy requirements in the cold by as much as 15% -35%.

B-15

b) Burrowing, Nest-Building and Migration.

All these behavioral responses protect endotherms from the effects of an extremely cold environment. They allow them to escape the cold by moving into a less extreme environment. Burrows serve two major functions: protection from predators and protection from extremes of ambient temperature. Nest-building further reduces the thermal conductance by using external insulation, i.e., nesting material. For example, lemmings with a nest of cotton wool can reduce their effective thermal conductance by 40%. Some endotherms escape the extreme cold by migrating into regions with a warmer climate.

c) Responses Facilitating Heat Gain.

Sun basking - Some endotherms, including humans, stretch out in the sun, to profit from heat gain by radiation (when the gain is larger than heat loss).

Hoarding of food, increase in food intake. During winter, food is scarce. Therefore, some animals increase food intake in autumn or hoard food in burrows. Depending on body size and food availability, some species increase heat production, or decrease it in winter (torpor, hibernation).

All these behavioral responses can be combined. Thus, an animal can increase its fat reserves in autumn, and escape the extreme cold by staying in a nest built in an under-soil burrow. In the nest, it can huddle together with others, or change its shape by curling up if alone, and reduce its metabolic rate by going into hibernation. Humans have only limited mechanisms to adapt to cold by autonomic means, which are first of all reserved for emergency. Normally, behavioral thermoregulation occurs before thermoregulatory heat production can increase. We dislike shivering and try to avoid it by using behavioral means. They may be as simple as reducing the conductance by adding clothing, hoarding food and fuel before the winter comes, hiding in a nest (the bed) made of furs of animals or in a snow burrow (an igloo), and making fire to heat food and drinking water. Of course, they may also be as complicated as using a thermoconditioned water bed in an air-conditioned apartment house, and flying by jet-plane to escape the cold season by spending the winter in a comfortable climate. We work to get the money to pay for all this thermal comfort! Thus, to conclude this chapter, endotherms differ in their size and insulation, and therefore use different thermoregulatory mechanisms and thermoadaptive strategies. These can be activated subsequently, depending on time factors and the rules of economy. Humans use their brain to design artificial burrows (for example, space bases), allowing them to expand into environments in which no other species can survive. This is, however, very expensive, a form of human behavior in extravagant opposition to the rules of economy.

Summary

Adaptive changes are complex, and include behavioral, functional and morphological modifications. The whole cold-adaptive process consists of several components that have different time requirements and overlap each other. They are governed by the rules of economy. The cold-adaptive strategy varies in different species according to their needs and the individual and inherited experience. This should be considered in comparative studies, and generalizations based on results in one or few species should be avoided.

Acknowledgments

I would like to express my gratitude to all those who assisted in the preparation of this chapter. Special recognition is due Erika Länger for editorial assistance and Birgit Störr for preparation of the figures. I am indebted to Dr. Joachim Roth for reading the chapter. His comments and criticism have enlighted and encouraged me.

References

The concepts used in this chapter are described in more detail in the following two monographs:
1. Hensel H, Brück K and Raths P (1973). Homeothermic organisms. In: Precht H, Christophersen J, Hensel H, Larcher W (eds), *Temperature and Life*, Springer, Berlin, pp. 503-779.
2. Withers PC (1992). *Comparative Animal Physiology*. Saunders College Publ., Fort Worth, pp. 1-949.

Two excellent collections of recent reviews to corresponding topics are recomended for details and original literature:
3. Schönbaum E, Lomax P, eds [1990]. *Thermoregulation A: Physiology and Biochemistry, B: Pathology, Pharmacology, and Therapy (Int. Encyclopedia of Pharmacology and Therapeutics)*. Pergamon Press, New York.
4. Fregly MJ, Blatteis CM, eds (1996), *Handbook of Physiology, Section 4: Environmental Physiology*, Volume I and II, Oxford University Press, New York.

Only few original papers are cited in the text:
5. Scholander PF, Hock R, Walters V, Johnson F and Irving L (1950). Heat regulation in some arctic and tropical mammals and birds. *Biol. Bull.* **99**: 237-258.
6. Brück K (1978). Heat production and temperature regulation. In: Stave U (ed), *Perinatal Physiology*. Plenum Publ. Corp., New York, pp. 455-498.
7. Heldmaier G (1971). Zitterfreie Wärmebildung und Körpergröße bei Säugetieren. *Z.Vergl.Physiol.* **73**: 222-248.
8. Zeisberger E (1997). Biogenic amines and thermoregulatory changes. In: Sharma HS, Westman J (eds), Brain Function in Hot Environment: Basic and Clinical Perspectives. Elsevier Science Publ., Amsterdam, in press.
9. Zeisberger E and Ewen K (1983). Ontogenetic, thermoadaptive and pharmacological changes in threshold and slope for thermoregulatory reactions in the guinea-pig (Cavia aperea porcellus). *J.therm. Biol.* **8**: 55-57.

10. Jöckel J and Zeisberger E (1994). Thermoregulatory changes after neurochemical lesions of catecholaminergic and serotonergic neurons in the lower brain stem of the guinea pig. In: Zeisberger E, Schönbaum E, Lomax P (eds), *Thermal Balance in Health and Diseases*. Advances in Pharmacological Sciences, Birkhäuser Verlag, Basel, pp. 73-78.
11. Sehic E, Ungar AL, Blatteis CM (1996). Interaction between norepinephrine and prostaglandin E_2 in the preoptic area of guinea pigs. *Am.J.Physiol.* **271**: R528-R536.
12. Sessler DI (1994). Thermoregulation and heat balance: general anesthesia. In: Zeisberger E, Schönbaum E, Lomax P (eds), *Thermal Balance in Health and Diseases*. Advances in Pharmacological Sciences, Birkhäuser Verlag, Basel, pp. 251-265.

Self-Study Questions

1. What is the difference between cold adaptation, acclimation, and acclimatization?
2. Define the range of tolerance, the range of resistance, the range of regulation, the thermoneutral zone, and the interthreshold zone?
3. What are the influences on heat loss of body size and insulation?
4. What are the main strategies for survival in the cold?
5. What factors can decrease heat loss?
6. What mechanisms contribute to the redistribution of blood flow in the cold?
7. What is the role of arteriovenous anastomoses in the cold?
8. What limits evaporative heat loss in the cold?
9. What mechanisms increase heat production in the cold?
10. What is heterothermy, and how is it differentiated?
11. How is the thermoregulatory system organized?
12. How can the inputs from sensors to the controller be modified, and how can the integration of thermal signals in the hypothalamus be modified?
13. What is the sequence of adaptive components in the cold adaptive process, and what is the role of behavior in this process?

Chapter 12.1

Learning Objectives
1. Definition of heat illnesses.
2. Basic understanding of the pathophysiology of heat injury
    Cellular components
    Integrative components.
3. Understanding of the factors predisposing to heat illness.
4. Understanding of the factors decreasing susceptibility to heat illnesses.

Bullets
B-1 Heat illness – How to define.
B-2 Heat stroke – A critical factor is the inability to sustain circulatory function.
B-3 Decreased plasma volume as a basis for the development of heat stroke.
B-4 Do splanchnic perfusion, endotoxins and lactacidemia play a role?
B-5 An ultimate, irreversible injury, is the cellular injury!
B-6 Tolerance to hot environments is reduced and, therefore, susceptibility to heat illnesses is increased by a variety of physiological/environmental factors.
B-7 The ability to withstand heat stress is improved following heat acclimation and enhanced physical fitness.
B-8 The heat shock response is a rapid, powerful, molecular mechanism providing transient enhanced thermotolerance.
B-9 Heat illness: an integrative or cellular failure.

Chapter 12.1

PATHOPHYSIOLOGY OF HYPERTHERMIA

MICHAL HOROWITZ[1] AND J.R.S. HALES[2]
[1]*The Hebrew University, Hadassah Medical School Department of Physiology, 91120 Jerusalem, Israel*
[2]*The University of Sydney, Faculty of Veterinary Science Camden, N.S.W., Australia*

1. Introduction

Homeotherms control their body temperature within a very narrow temperature range of ca. 1°C. The thermoregulatory system, although "competing" with other homeostatic control systems which use the same effectors, operates on a very high gain (gain = corrected disturbance/remaining error). Therefore, when body temperature increases above the normal range (hyperthermia) because of environmental or metabolic heat loads, clinical failure (heat illness) may rapidly develop. Heat illness represents a spectrum or continuum of disorders ranging in severity from temporary and mild to fatal, depending not only on hyperthermia *per se* but on a variety of "host factors" which affect the magnitude and rate of rise of body temperature via their interference with temperature regulation or other related homeostatic control systems. This chapter defines the concept of heat illness and discusses the physiological and pathological factors predisposing to, and those attenuating, the detrimental effects of hyperthermia.

2. Heat Injuries

2.1. Heat illness - definitions

The following terms have been defined for clinical diagnosis. They can also be used to describe similar phenomena in experimental animals.

Heat strain covers all effects of excess body heat, but the term is commonly restricted to the earliest, mildest effects of which one might be barely conscious.

Heat edema, seen as swelling in the lower limbs, occurs in the unacclimatized state, during inactivity within 1-2 days of entering a hot climate; it may be aggravated or even induced by excessive salt intake, and is usually associated with oliguria secondary to heat-induced vasodilatation.

Heat syncope is the orthostatic dizziness or syncope which occurs in an upright person in a hot environment. It is generally attributed to reduced cerebral blood flow (BF), resulting from a decrease in arterial pressure due to poor venous return following heat-induced cutaneous vasodilatation and, possibly, postural pooling of blood.

Heat cramps seldom occur without exercise. They are strong, painful tetanic contractions of specific groups of extremity or abdominal muscles, usually in very active people who, while sweating profusely, drink to replace water but not the salt deficit. Slight breakdown of muscle tissue (rhabdomyolysis) may occur.

Heat exhaustion is the first condition in the continuum of heat strain which may develop to heat stroke; it is diagnosed when a person generally exercising heavily in the heat physically collapses in a hot environment. Usually, body temperature is elevated above 38°C, the skin is pale (low blood flow) and clammy (still sweating), and there is a feeling of weakness, possible dizziness, headache, vomiting, diarrhea, and abdominal and extremity muscular cramps. It is attributable to water and/or salt depletion, and circulatory abnormalities may occur.

Heat stroke represents the collapse of heat-regulating mechanisms, which results in an explosive rise in body temperature, usually to above 40.5°C (but as low as 39°C). A consistent symptom is mental confusion or delirium and even coma. It is fatal if not promptly recognized and treated. It may be *"classical"* (CHS), or *"exertional"* (EHS). CHS most commonly occurs in very young or old, sedentary people in poor health . Usually, skin BF and sweating are low, and there is severe respiratory alkalosis; but the other features seen in EHS are mild or uncommon. EHS most commonly occurs in healthy, physically fit people about 15-45 years old. Usually, skin BF is low and sweating persists; respiratory alkalosis is mild. There is a host of other symptoms: lactic acidosis, rhabdomyolysis, hyperuricemia, elevated ratio of creatinine to blood urea nitrogen, elevated creatine phosphokinase (CPK), aldolase, hypercalcemia, disseminated intravascular coagulation (DIC), hypoglycemia, and possibly acute renal failure. Time should not be wasted on precise diagnosis - if a person collapses in a hot environment or during exhaustive exercise, treatment for heat stroke must begin without delay.

3. Heat Stroke: An Integrative View
There is a multitude of factors leading to the development of the heat stroke syndrome. We can distinguish between physiological inadequacy and effector malfunction. Concomitantly, the unavoidable imposition of a variety of stressors (predisposing factors), physiological and pathological, either accelerates or delays its development, depending on their mode and level of interference with body function.

The hyperthermic level at which heat stroke develops is also variable. Nevertheless the ultimate cascade of pathophysiological consequences is similar in all cases. Apparently, a critical factor is the inability to sustain circulatory function, as discussed below.

3.1. Integrative mechanisms underlying physiological failure during heat stress
The thermoregulatory system relies critically on adjustments in cardiovascular activity to transport heat throughout the body; in particular, as all heat loss must take place at the body surface (skin and nasobuccal tissues), increased blood flow to the body surface is imperative (see also chap.11.1). However, the cardiovascular system must provide a level of perfusion of all tissues which is adequate to meet

their metabolic needs; for this, arterial pressure must be maintained within a narrow range.

During mild heat stress, body surface blood flow is elevated without detrimentally interfering with the blood supply of other tissues, either via an increase in cardiac output (CO) and/or by a redistribution of the CO. For the latter, BF is taken away from what can be regarded as non-vital tissues or tissues that usually have a BF in excess of what they need to keep functioning adequately, viz., principally from the splanchnic area and probably also a little from kidneys, muscle and fat. However, it is apparent that when heat stress becomes severe, the demands for body temperature regulation may ultimately conflict with the requirements for blood pressure regulation. Evidence to date strongly suggests that such a conflict in homeostatic drives is the most important factor leading to the physiological breakdown (or the pathophysiological state), i.e., leading to the heat stroke syndrome. As discussed in Sec., 3.5, a conflict also develops within one system, viz., the thermoregulatory system itself.

3.1.1. The cardiovascular basis of heat stroke

Fig. 1 presents a scheme which has been synthesized to illustrate the proposed links between experimentally or clinically observed events occurring during heat stress in a person resting in a hot environment. The upper block describes the thermoregulatory phase when body temperature remains unchanged. With increasing heat load (Fig. 1, block 2), further increases in skin BF together with profuse sweating lead to increased peripheral blood volume, which together with decreasing total blood volume (due to sweat loss), deplete central blood volume. This reduces central venous pressure (filling pressure of the heart), which (presumably via low pressure baroreceptors in the cardiopulmonary regions) is soon followed by a diminution in the high levels of skin BF and volume, and hence, in sweating. Heat loss is thereby reduced and the development of hyperthermia is aggravated. That is, whereas priority was initially given to the thermoregulatory needs for high skin BF, at the more advanced stages of heat stress when maintenance of blood pressure becomes threatened, cardiovascular needs ultimately take priority over body temperature control. The high level of body hyperthermia *per se* and the detrimental cellular/molecular changes (Fig. 1, block 3) ultimately lead to the potentially fatal cascade of events consisting of acidosis (3.2.2), cerebral hypoxia, neural dysfunctions (3.3), disseminated intravascular coagulation (DIC), coagulative necrosis, and cell death (3.5), which, ultimately, lead to death of the organism.

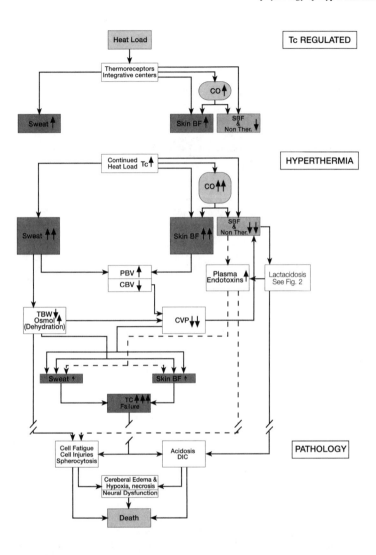

Figure 1. A proposed sequence of events leading to the development of heat stroke syndrome (in humans) during exposure to high ambient temperature under sedentary conditions. Box size and arrow numbers in boxes assigned for thermoregulatory effectors signify magnitude of change from normothermic level. Arrows indicate increase (↑) or decrease (↓) above normothermic level. Broken lines represent the presumptive role of endotoxins in the development of heat stroke (2.4.). BF - Blood flow, CO - Cardiac output, CBV - central blood volume, CVP - central venous pressure, DIC - disseminated intravascular coagulation, Osmol - plasma osmolality, PBV - peripheral blood volume, non-Ther.- Non thermoregulatory organs; SPF - splanchnic blood flow, TBW - total body water, Tc - core temperature. Adapted from Hales et al., 1996, with permission.

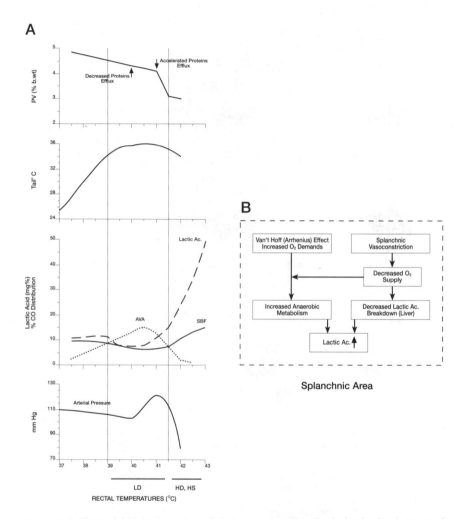

Figure 2 **A.** Thermal dehydration - decreased plasma volume as a basis for the development of heat stroke syndrome in the rat model. As long as plasma volume (PV) is conserved adequate vasomotor thermoregulatory activity (tail and arteriovenous anastomoses [AVA] blood flow) is maintained. The failure to conserve plasma water leads to breakdown of the vasomotor heat dissipation mechanisms. An abrupt rise in plasma lactic acid concentration accelerates this detrimental effect. A marked rise in systemic blood pressure precedes heat stroke. LD - Low dehydration, HD - High dehydration, HS - Heat stroke. In **B.** the cascade of events leading to lactacidosis is presented. Adapted from Horowitz and Samueloff, 1987, 1988, and Zurovski et al., 1991, with permission.

3.1.2. Thermal dehydration decreased plasma volume as a basis for the development of heat stroke

The conservation of plasma volume is a major determinant of the cardiovascular capacity influencing endurance in a hot environment. Thermoregulatory failure is coincident with abrupt loss of plasma volume and often precedes circulatory failure. The interrelationship between plasma volume and the circulation is clearly demonstrated in the thermal dehydration-heat stroke experimental model. That is, during heat stress, if no water is available, the intense activation of the evaporative cooling system leads rapidly to dehydration. This is because a disproportionate loss of plasma vis-a-vis total body water develops due to the lag in the shift of water from the intracellular to the extracellular compartments to replace the water lost from the plasma for evaporation. This loss, superimposed on the thermally induced redistribution of plasma volume (see 3.1.1), limits the capacity of the circulating blood to convect heat from the core to the periphery, and promotes the circulatory failure-induced heat stroke syndrome (Fig. 2). During mild to moderate dehydration, efficient operation of plasma volume conserving mechanism (for details see chap. 11.1) attenuates this deterioration, thus allowing an augmented skin BF, including perfusion of arteriovenous anastomoses (AVA), via a shift of blood out of the splanchnic and skeletal muscle vascular beds. With progressive dehydration, closure of AVA, a manifestation of thermoregulatory failure, and resumption or even elevation of splanchnic BF occur, concomitantly with an increase in plasma protein extravasation, accompanied by further water loss from the circulation and an abrupt rise in body temperature. These events also cause an abrupt lactacidosis (see 3.2). There is a temperature threshold (approximately 41-41.5°C) at which lactic acid (LA) starts to rise abruptly. LA is a potent vasodilator and appears to contribute/accelerate the failure of plasma volume conservation mechanisms and the subsequent thermally induced circulatory failure. At extremely high ambient temperatures, the causal relationships between lactacidosis and a significant reduction in plasma volume occur even at a very minor level of dehydration; this substantiates the lactate contribution to the development of heat stroke.

B-3

3.2. Splanchnic perfusion and metabolic acidosis

Metabolic acidosis due to increased LA concentration, combined with high arterial PO_2 is a common observation during hyperthermia. In addition to its detrimental effect on plasma volume, it not only produces a negative inotropic effect on the heart and interferes with potassium balance, but also denotes cellular fatigue (3.5). Under sedentary conditions, the occurrence of lactacidosis in the face of adequate circulatory PO_2 may suggest that LA is of local origin. From thermoregulatory considerations, we may hypothesize that the major source of LA is splanchnic tissue where the thermally induced reduction in circulating blood supply brings about an accelerated localized hyperthermia and, in turn, increased oxygen

B-4

demands (van't Hoff effect), combined with reduced oxygen supply. These are met by a shift to anaerobic metabolism and increased production of LA. Impairment of LA turnover in the liver has also been reported.

3.3. Circulation induced cerebral edema and brain dysfunction

As noted in 3.1.5., brain dysfunction is regarded as a key diagnostic marker for heat stroke, yet there have been few investigations of this. Cerebral edema and elevated intracranial pressure are common observations during heat stroke. However, it is not yet conclusive whether cerebral ischemia or cellular fatigue (cytotoxic edema) leads to this pathological response. In rabbits, arterial hypotension associated with cerebral ischemia is coincident with the onset of heat stroke, and elevated cerebral BF (by deafferentation of the superior cervical ganglion) delays heat stroke. In rats, accelerated protein extravasation suggests circulatory failure-induced edema as well. Brain dysfunction has been studied experimentally by measuring auditory and visual evoked potentials. As body temperature increases, the latencies of the waves decrease. It seems that the greatest latency decrease occurs at the more rostral levels of the brain stem.. During cooling, latencies resume their control level. However, amplitudes of the ABR (auditory brain stem) waves remained reduced, suggesting that upon return from the pathological state, fewer fibers are responding. Similar findings were described both for animal species and humans.

3.4. A role for endotoxins

Endotoxins normally raise body temperatures (cause fever) by raising the hypothalamic set-point; this leads to attenuated heat loss mechanisms and/or increased metabolic heat production. Recently, experimental evidence has indicated that endotoxins, which are normally present in large quantities in the gastrointestinal tract (GIT), can reduce heat tolerance. For instance, they are at least partly responsible for the relatively low heat tolerance of sedentary compared with physically fit people and in old age.

The gram-negative bacteria in the GIT shed their walls containing lipopolysaccharides (LPS) into their environment. The LPS is normally restricted to the GIT by the extremely effective barrier of the gut wall; however, plasma levels of LPS increase during heat stress for two reasons: (a) the barrier to movement from the GIT into the blood system is diminished by the reduction in GIT BF (Fig. 1) (and probably also by hyperthermia *per se* and by other factors such as acidosis and hypoxia); (b) the reticuloendothelial system, which normally removes circulating endotoxins, is reduced in efficiency in the hyperthermic state.

In Sec 3.1.1, we have seen how, during severe heat stress, a conflict develops *between* autonomic mechanisms, viz., body temperature vs. blood pressure control. Here, with respect to endotoxins, a conflict develops *within* the thermoregulatory system, viz., stimulation of heat loss responses to combat a rising body temperature (due to the hot environment) vs. inhibition of heat loss responses to raise body temperature (to develop a fever to combat the endotoxins). Thus, note in Fig. 1 that besides the cardiovascular changes discussed in Sec. 3.1.1., the decrease in GIT BF (included in SBF) leads to increased plasma endotoxin levels. It is probable, although not yet proven, that this is at least partly responsible for the reductions in sweating rate and skin BF and for some of the hematological changes which are seen in advancing stages of severe heat stress. In addition to fever, endotoxins enhance DIC and can severely compromise myocardial function. Many features of heat stroke resemble endotoxin shock, viz., DIC, petecheal hemorrhages, diarrhea, hyperkalemia, and often hypotension and acidosis.

3.5. Cellular components of heat injury

The cellular components of heat injury include an enormous number of effects, including inhibition of protein synthesis, cytoskeleton damage, acidosis, altered signal transduction pathways, elevated cytosolic Ca^{2+}, cell and mitochondrial swelling, etc. In fact, these are unspecific responses to many stressors (e.g. hemorrhage, sepsis, ischemia). However, careful examination of the factors leading to heat illnesses suggests that cellular fatigue, due to energy depletion, is critical. It is suggested that, during hyperthermia, cells, by coping with leaky membranes, accelerate exothermic Na^+/K^+ ATPase, thereby producing even more heat, while attempting to elevate their metabolic level. This positive feedback leads to a reduced steady-state energy level, cell fatigue and acidosis, and is aggravated by the effect of temperature on membrane fluidity. Membrane lipids form an important solvent compartment for membrane-bound proteins (pumps, e.g. Na^+/K^+ ATPase, membranal receptors, G-proteins, etc.). Hence, changes in the lipids' physical state or composition may modulate protein activity, e.g., increasing sodium transport and, in turn, the Na^+/H^+ antiport-exchanger, leading to intracellular acidosis. Parts of this hypothesis have been proved, and to many investigators (and clinicians) the cell membrane is the first cellular target of heat injury.

B-5

An additional cellular phenomenon, repeatedly observed upon heat stroke, but as of yet undetermined significance, is the change of a large proportion of erythrocytes from their normal biconcave discoid to spherical shape. Such "spherocytes" not only increase blood viscosity and enhance DIC, but they are unlikely to be able to deform to the parachute shape which is normally necessary for erythrocytes to squeeze through the microvasculature, and they could therefore form microemboli. Physical fitness largely avoids spherocytosis.

4. Factors Predisposing to Heat Illness

Tolerance of hot environments is reduced and, therefore, susceptibility to heat illnesses is increased by a variety of physiological/environmental factors, each interfering differently with thermal balance.

4.1. Exercise

Exercise reduces heat tolerance in several ways. (a) Muscle contraction is a highly inefficient process, with approximately 80% of the chemical energy used being converted to heat. Up to ca. 20x basal heat production can be added by exercise, and almost all of that heat must be carried away from the muscles by the blood stream to the core and, then, to the body surface. (b) There is a greatly increased load on the cardiovascular system, partly representing a third homeostatic drive competing with the regulation of blood pressure and body temperature. The increased metabolism of active muscles demands an increased blood supply. In fact, with the combined loads of exercising in a hot environment, both muscle BF and skin BF, and therefore exercise and thermoregulatory capacities, respectively, are compromised. Thus, at the same level of exercise, muscle BF is lower in a hot than in a thermoneutral environment; on the other hand, at the same level of environmental heat load, skin BF is lower during exercise than at rest. (c) Plasma volume is reduced due to an isotonic shift of plasma fluids into the active muscles. (d) There is increased cellular fatigue due to accelerated energy drain and augmented acidosis (see also 3.5). These events could, in part, explain the higher morbidity and mortality observed in exertional vs non-exertional heat stroke.

4.2. Hypohydration

Among the physiological "host factors" accelerating hyperthermia, hypohydration is the most common. Hypohydration is usually characterized by chronic, lowered total body water with or without hyperosmolarity. Under these conditions, thermoregulation is altered. Lowering of heat production is coincident with an increase in basal and threshold temperatures for the activation of heat dissipating mechanisms. This leads to improved utilization of the physical avenues for heat dissipation. The heat stroke syndrome is thus developed at a higher body temperature than during euhydration. However, due to the lowered volume of the circulating blood and of total body water, the capacity of thermoregulatory effectors decreases and heat endurance is significantly reduced. This condition is severely aggravated if exercise is added. In humans, a ca. 3% loss of body weight produces detrimental effects, such as reduced sweating and reduced skin BF.

"Voluntary dehydration" is the name given to the phenomenon in which a person's thirst drive is insufficient to cause drinking of sufficient quantities of water to fully replace the body water deficit. This phenomenon appears to be due to loss of ions, and, therefore, restitution of body water following hypohydration is likely to be complete only if the lost ions are replaced.

4.3. Pregnancy

Pregnancy decreases heat tolerance, partly because of a slight elevation in heat production, but principally because of its cardiovascular requirements. Blood volume increases with pregnancy; however, the needs of the pregnant uterus for BF compete successfully with the BF requirements for cutaneous vasodilatation and of exercising muscle.

4.4. Aging

Invariably, during a heat wave, the greatest morbidity and mortality occur in the elderly (see also Chapt. 9.2). It appears that neither sensitivity nor capacity for sweating is diminished. In contrast, cardiovascular performance is very sluggish and reduces heat loss efficiency. This would conform with the decline in maximal oxygen uptake capacity which occurs with increasing age (above ca. 20 yrs.). Recent evidence indicates that endotoxins contribute significantly to the lowering of heat tolerance (Sec. 3.4) in old age.

In children, the higher ratio of surface area to mass provides a fundamental characteristic which enables greater heat exchange - either gain or loss. However, the latter does not occur well in the heat, not only because maximal sweating capacity takes many years to develop, but, more importantly, because of the limited cardiovascular reserves of the young, growing body. For example, resting heart rates are higher in children, and, since the maximum is roughly independent of age, their capacity for increase is limited. Further disadvantages for children are a higher metabolic rate per unit mass and a slower rate of acclimatization.

4.5. Drugs

Drugs that aggravate hyperthemia fall into five categories, viz., those causing muscular hyperactivity, causing hypermetabolism, impairing central thermoregulatory mechanisms, impairing thermolytic mechanisms, or impairing cardiovascular compensation. For example, anticholinergic drugs (such as atropine, used for protection in chemical warfare) inhibit sweating, and, therefore, seriously reduce heat tolerance. Alcohol consumption promotes heat exhaustion, presumably due to shifts between body fluid compartments.

4.6. Preexisting illnesses

Preexisting illnesses which have any of the effects listed above will reduce heat tolerance. For example, any heart or vascular condition which compromises increases in cardiac output or regional variations in vascular tone will restrict a person's capacity to lose heat. Diabetics appear to be sensitive to heat, which could be related to their commonly having faulty baroreflexes (easily exhibiting postural hypotension) or sweating abnormalities, and neuropathies of the autonomic nervous system.

5. Factors Decreasing Susceptibility to Heat Illnesses
The ability to withstand heat stress is improved and the susceptibility to heat illnesses is decreased by all physiological factors which improve thermal balance, decrease heat production, and enhance thermotolerance.

5.1. Heat acclimation
A comprehensive discussion of the beneficial effect of heat acclimation and the mode of adaptation of the thermoregulatory effectors is given in Chap. 11.1. Heat acclimation enhances heat tolerance and, in turn, delays the development of heat stroke, basically by expanding the thermoregulatory range, as follows: 1. expansion of temperature safety margins; 2. enhancement of the capacity of the thermal effectors; and 3. decreased heat production. Among these three factors, the first is still somewhat puzzling. While decreased basal core temperature is unequivocal, our knowledge of the level of hyperthermia at which heat stroke develops is rather confusing. In various species and under different conditions, heat acclimation reduces, elevates, or does not change the lethal temperature. Differences might be associated with heat dose and rate of heating (see also 5.3). Collectively, however, the temperature safety margin is expanded. Concomitantly, the temperature thresholds for activation of the evaporative cooling system and skin vasodilatation decrease. An important contribution of the cardiovascular system to the delayed development of heat stroke is the augmentation of splanchnic blood flow upon heat stress. Together these adaptive changes allow adequate skin perfusion and improved splanchnic blood flow, hence better heat dissipation from the body core while maintaining the endotoxin barriers. Improved utilization of the intracellular water reservoir to support plasma volume conservation, and sometimes augmentation of plasma volume also occur.

Whether heat acclimation affects the detrimental cascade *per se* is unclear. In studies carried out on rats, it was shown that heat acclimation blunts the failure of the splanchnic vasoconstriction reflex (possibly due to reduced nitric oxide production)

An important effect of heat acclimation is attenuation of the aggravating effect of exercise. In humans, the severe hyperthermia of exercise is delayed by heat acclimation. A study on rats, designed to assess the separate contributions of heat and exercise to improved performance in a hot environment, showed that the share of heat acclimation is greater than that of exercise.

5.2. Physical fitness
The greater cardiovascular capacity of the physically fit not only enhances their available, thermoregulatory increase in skin BF, but, probably more importantly, requires less reduction in their GIT BF; endotoxins, therefore, play a more significant role in lowering the heat tolerance of sedentary than of fit people. In

addition, physical training lowers the threshold temperatures for onset, and increases the maximum capacities of both sweating and cutaneous vasodilatation, thus increasing heat tolerance.

5.3. Heat shock response

The heat shock response (HSR) is a rapid, but transient (several days) adaptive mechanism which provides thermotolerance (improved ability to sustain higher temperatures in the future). It is elicited following heat stress. Cellular and molecular processes start within minutes. However, in the whole, intact body, HSR is fully expressed approximately twenty-four hr after the given stress by a significant increase in the ability to withstand heat stress (+60% in rats and mice) and higher body temperatures (thermotolerance). The mechanisms providing this protective effect involve both cellular and integrative components. The cellular and molecular mechanisms of this process are better understood than the integrative ones. The effects appear to be controlled at the molecular level, representing a "cellular attempt" to protect vital, ultrastructural components from thermal damage in a way that facilitates survival and subsequent recovery after the stress is removed. Principally, there is synthesis of highly conserved families of proteins, the heat shock proteins (HSP), which are molecular chaperons that bind to denatured or nascent proteins to prevent their unfolding. Among the HSPs, the 70 kD family is the most responsive to heat stress. Their cytoplasmic level correlates with the magnitude, duration, and rate of heating. Their potential utilization as a biomarker of organ/tissue heat injury was recently examined. The results were encouraging. Cells subjected to heat shock may also induce HSP-independent protective mechanisms, e.g., increased synthesis of glycoproteins, accumulation of polyols and sugars, and increased synthesis of antioxidant enzymes. Less familiar, although apparently indispensible, are the HSR mechanisms leading to improved systemic functions, e.g., enhanced mechanical performance of the heart and the vasculature. Considering the timing of its full expression, HSR might be considered as the first line of cellular defense, and it is likely to have beneficial effects during periodic heat waves or frequent exposures to heat stress.

B-8

6. Summary: Integrative vs Cellular Failure

Heat illness represents a continuum of disorders ranging in severity from temporary and mild to fatal. The breakdown step leading to severe hyperthermia is the inability to sustain circulatory function. Core temperatures of 40°-42°C are commonly considered dangerous. These temperatures are usually associated with brain dysfunction; to date, the data are insufficient to conclude whether synaptic transmission, conductivity, or cell injury *per se* is responsible for this damage. General tissue damage is also observed in that temperature range. The individual cells, however, can sometimes sustain temperatures higher than those recorded during heat stroke, There is also variability in thermal sensitivity (for both

B-9

temperature and heating rate).of various cells. It is likely that the concerted factors developed during hyperthermia evoke cellular injury; they may also decrease the temperature threshold for cell death. It thus appears that heat stroke is a systems rather than a cellular failure.

7. Suggested Readings

1. Hales JRS (1987) Proposed mechanisms underlying heat stroke. In: Hales JRS, Richards DAB (eds) *Heat Stress: Physical Exertion and Environment.* Elsevier, Amsterdam, pp. 85-102.

2. Hales JRS, Hubbard RW, Gaffin SL (1996). Limitation of heat tolerance. In: Fregly MJ, Blatteis CM (eds), APS Handbook of Physiology, See. 4, *Environmental Physiology.* Vol. 1. Oxford University Press, New York - Oxford, Chap. 15.

3. Horowitz M (1984) Thermal dehydration and plasma volume regulation: Mechanism and control. In: Hales JRS (ed) *Thermal Physiology.* Raven Press New York, pp.389-395.

4. Horowitz M, Samueloff S (1988) Cardiac output distribution in thermally dehydrated rodents. *Am. J. Physiol.* 254:R109-R116.

5. Hubbard RW (1990). Heatstroke pathophysiology: the energy depletion model. *Med. Sci. Sports Exerc.* 22:19-28.

6. Knochel JP (1974). Environmental heat illness: An eclectic review. *Arch. Intern. Med.* 133:841-864.

7. Zurovski Y, Eckstein L, Horowitz M (1991). Heat stress and thermal dehydration: Lactacidemia and plasma volume regulation. *J. Appl. Physiol. 71*: 2434-2438.

8. Self Evaluation Questions

1. What is the definition of heat illness?

2. What is the difference between heat exhaustion and heat stroke?

3. What is the breakdown step leading to the development of the heat stroke syndrome?

4. What is the role played by plasma volume in the development of the heat stroke syndrome?

5. How does lactacidemia develop?

6. What is the mechanism of action of endotoxins?

7. What are the cellular manifestations of heat injury?

8 What is the common denominator of all factors increasing and of all factors decreasing susceptibility to heat illness?

9. What are the mechanisms of action of heat shock proteins?

Chapter 12.2

Learning Objectives
1. To be able to define the different types of hypothermia with respect to severity and duration.
2. To be able to describe the various physiological responses to cold exposure and how these change with age.
3. To be able to distinguish between freezing and non-freezing cold injuries.
4. To recognise that exposure to cold may lead to thrombogenesis, particularly in the elderly.

Bullets
B-1   Humans are tropical animals
B-2   Cold water
B-3   Wind chill
B-4   AVA's
B-5   Fibrillation
B-6   Nerve conduction
B-7   Cold-induced vasodilation
B-8   Impaired thermoregulation
B-9   Thrombosis
B-10 Cold weather
B-11 Housing standard

Chapter 12.2

HYPOTHERMIA AND COLD INJURIES IN MAN

JAMES B. MERCER
*Department of Medical Physiology, Institute of Medical Biology
University of Tromsoe, N-9006 Tromsoe, Norway*

1. Introduction

In homeotherms, including humans, body core temperature is one of the most vital, constant, regulated parameters, the normal regulated level of body core temperature being referred to as *normothermia*. From a thermal physiological point of view, humans are tropical animals when naked, being most comfortable at an ambient temperature of 28 -30 °C, the so called *lower critical temperature* (LCT). At ambient temperatures below the TLC, humans need to invoke *heat defence mechanisms*, including reducing heat loss by vasomotor control or increasing heat production metabolically (see below) in order to maintain a state of normothermia. In this chapter, we will examine the effect of being exposed to ambient temperatures below the lower critical temperature, which may lead to a lowered body core temperature - *hypothermia*. We will discuss and define different types of hypothermia, both with respect to the degree and duration of body cooling as well as the physiological responses. In addition, we will discuss other effect of cold, including freezing and non-freezing cold injuries as well as the effect of alcohol during cold exposure. Finally, since most countries report large increases in mortality due to cardiovascular disease during the colder winter months, we will discuss this in relationship to age-related changes in the ability to control body temperature as well as cold-induced changes in the haemostatic system.

2. Hypothermia

Hypothermia is generally defined as a core (rectal, oesophageal, or tympanic) temperature of less than 35°C (95°F). Core body temperatures in the range 34°C to 35°C (93.2°F to 95°F) are considered mildly hypothermic; 30°C to 34°C (86°F to 93.2°F), moderately hypothermic; and less than 30°C (86°F), severely hypothermic. *Acute or accidental hypothermia* may be defined as an accidental or deliberate decrease in core temperature, usually in a cold environment and associated with an acute problem, but without primary pathology of the temperature regulating system. The various degrees of hypothermia can also be defined in terms of duration: *acute hypothermia* (few hours), *prolonged hypothermia* (several hours), and *chronic hypothermia* (days or weeks). Since water has a specific heat, thermal conductivity, and density 1000 times, 25 times, and over 800 times more than air, respectively, body core cooling is much more rapid during cold water immersion *(immersion hypothermia)*. In addition to hypothermia resulting from environmental exposure (cold air stress, cold water immersion) or by induction for surgical purposes, a reduction of core temperature can also be precipitated by a number of other conditions, including metabolic disorders (e.g., hypothyroidism, hypoglycaemia, malnutrition), central nervous system (hypothalamic) disorders, sepsis, drugs (notably barbiturates, phenothiazines), alcohol abuse, extremes of age, as well as attenuated behavioural thermoregulation. *Occupational hypothermia* describes a hypothermic state resulting from working in cold environments and arises from a conflict between measures for the maintenance of body heat balance and measures for the maintenance of work efficiency and performance. Measures for the maintenance of heat balance tend to increase the energy cost of work, decrease mobility and dexterity, and reduce efficiency and productivity.

The most evident and common hazard in the cold is *wind chill*, and the wind chill index (WCI) has been used as a means of interpreting the hazards of low temperature and wind. Derived from laboratory studies on water-heated cylinders, the WCI is a measure of the estimated heat loss per unit surface area ($W \cdot m^{-2}$) from exposed, unprotected surfaces of the human body and is calculated by the following equation

$$WCI = 1.16 \times (10.45 + 10\sqrt{v_{ar}} - v_{ar})(33 - T_a) \qquad [1]$$

where v_{ar} is the relative air velocity in $m \cdot s^{-1}$ and T_a is the ambient temperature in °C. Exposed parts of the body (e.g., face) are particularly susceptible to the cooling effect of wind. At a T_a 20°C and a wind speed of 4 $m \cdot sec^{-1}$, the WCI is 380 $W \cdot m^{-2}$. At a WCI of 1400 $W \cdot m^{-2}$, flesh will begin to freeze, and at 2300 $W \cdot m^{-2}$, exposed facial areas will freeze within a minute. A second hazard is *extremity cooling*. The maintenance of warm hands and feet is essential to human function and survival in the cold. The large surface area to mass ratio of the extremities, particularly the fingers, makes them susceptible to rapid cooling. Due to the presence of *arterio-venous anastomoses (AVA's)*, circulation can be shortcut through the fingers. In this way, blood flow rate can be maintained at very high levels, allowing a considerable heat input to the hand. The function of the AVA's is under fine control of the thermoregulatory centre and dependent on the general thermal status of the body (see also Chap. 5 and 11.2). Avoidance of extremity cooling is largely a question of maintaining a positive central heat balance while keeping the AVA's open. Once the AVA's have closed, the extremities will cool at a rate depending mainly on the rate of physical heat exchange. Finally, when there is negative balance between body heat production and heat loss, *whole body cooling* will occur.

B-3

B-4

3. Pathophysiological Effects of Cold

3.1. Heat production and heat loss

Hypothermia affects all organ systems, and the physiological responses to hypothermia are many and varied. Fig. 1 describes the main symptoms associated with falling body temperature during immersion in cold water, while Table 1, from a review by Granberg [2], summarises the general affects of hypothermia on different organ systems. As a first line of defence, heat loss is minimized by *peripheral vasoconstriction,* an effect which is most pronounced in the extremities. In an adult human, changes in the rate of blood flow through skin and extremities can change the *thermal conductance* by a factor of 4 to 7, depending on the thickness of the body shell and of the subcutaneous fat. Vasoconstriction does not occur in the head-neck region. The second main line of defence is *shivering thermogenesis,* in which *metabolic rate* may be increased by as much as 6-fold (see also Chap. 4.1). *Non-shivering thermogenesis* associated with active *brown adipose tissue (BAT)* occurs in the newborn (including humans) and in small cold-adapted animals (see also Chap. 4.2).

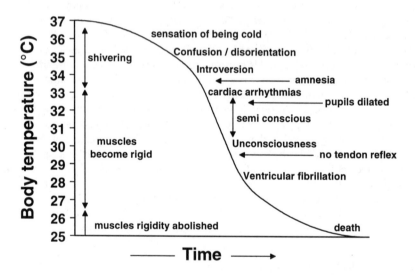

Fig. 1. Symptoms associated with falling body temperature during immersion in cold water. Redrawn after Golden [1].

Table 1. Effects of hypothermia on different organ systems,

Cardiovascular System	°C	Respiratory System	°C	Renal Function
Peripheral vasoconstriction	37	↑Resp. rate; resp. alkalosis bronchospasm	37	Cold diuresis (x 2 - 2.5)
↑HR ↑ BP ↑ CO	36	Immersion: initial gasp + + hypervent. (x5)-hypocapnia	36	RBF & GFR →
BP difficult to obtain J-wave	34	Respiration shallow Bronchorrhea	34	FF slightly ↑
Prolonged systole 30% ↓ HR & CO Atrial & ventricular dysrytmias 50% ↓ HR & CO	32	50% ↓ CO_2-production Respiratory acidosis 50% ↓ O_2-consumption	32	Extraction ratio of PAH → 50% ↓ RBF Autoregulation still intact
Death due to ventricular fibrillation or asystole (Cave! Shivering can mimic ventricular fibrillation!)	28		28	RBF & GFR ↓ parallel to ↓ CO
	26	Pulmonary oedema Apnea may be present	26	
	24	75%↓ O_2-consumption	24	

Central nervous system	°C	Muscles & peripheral nerves	°C	Endocrine systems & metabolism
Tiredness		↑↑ Symp. nervous discharge		↑↑NA ↑corticoids ↑ ALDO
Silence Apathy	36	↑ tendon reflex Dysarthria Fumbling & stumbling	36	↑T_4 ↑ T, TSH → Shivering
Slow cerebration poor judgement Retrograde amnesia	34	Ataxia	34	Non shivering thermogenesis? Maximal shivering
	32	Muscle rigidity	32	
Consciousness often clouded Stuporous; hallucinations Paradoxical undressing Pupils dilated	30		30	Insulin inactive Shivering disappears 50% ↓ Basal metabolic rate
No muscle reflexes No response to pain No pupillary light reflex	28	25%↓ Nerve conductivity	28	Failing heat conservation Poikilothermia
	26	Physical activity impossible	26	
No corneal reflex Cerebro-vascular autoregulation disturbed	24		24	
Flat EEG at 19°C	22	Local temperature 12°C crit. max. for manual dext. 10°C " " " tactile sensitivity 8°C complete nervous block	22	80%↓ Basal metabolic rate

Reproduced from Granberg, PO [2] with permission from the publishers.

3.2. Hypoglycaemia

With decreasing temperature, blood glucose levels increase and insulin levels decrease. However, *hypoglycaemia* may also occur and is often a result of glycogen depletion caused by shivering. Insulin is inactive below 31°C (88°F). At 26°C (79°F), the human body fails to conserve heat and becomes *poikilothermic*. In the absence of shivering, metabolism decreases by about 8% per 1°C reduction in body core temperature, to one-half of normal at 28°C (82°F).

3.3. The cold heart

Patients with moderate (34-30°C; 93.2 - 86°F) or deep (<30°C; <86°F) hypothermia exhibit characteristic changes in cardiac function. Early tachycardia and elevated blood pressure, followed by progressive *bradycardia* and lowered pressure, are common cardiovascular effects of hypothermia. With progressive cooling, there is a risk of *atrial fibrillation, ventricular extrasystoles, ventricular fibrillation* (28 - 25°C; 82 - 77°F), or *cardiac arrest* and *global ischaemia*. During hypothermia, the ECG shows bradycardia, widening of the QRS-complex, prolongation of the QT-interval, and quite often an extra deflection at the QRS-ST junction called the *"J" wave* (or the Osborne wave). Many of these changes can be attributed to the temperature-dependence of the kinetics and conductance of ion channels responsible for the different currents underlying the cardiac action potential. During *rewarming*, there is a danger of cardiovascular arrhythmias, pulmonary edema and reduced left ventricular performance.

3.4. Cold-induced changes in the haemostatic system

There are several changes in blood composition known to occur during cold exposure. These include increased red cell count and increased plasma cholesterol; the latter which is responsible for platelet aggregation and increased plasma fibrinogen, which is strongly thrombogenic. During the coldest part of the year, there is a range of risk factors which may change, including blood pressure, lipids, platelets and other related thrombogenic factors. For example, it was been reported that, during winter, blood pressure may be elevated by 3-5 mm Hg, the concentration of high density lipoprotein cholesterol lowered by 0.08 mmol/l, fibrinogen elevated by 0.34 g/l, and α_2-macroglobulin, a protein that inhibits fibrinolysis, raised. Recent findings in elderly have also shown significant winter elevations in plasma fibrinogen as well as factor VII clotting activity.

3.5. Osmotic, respiratory and mental functions

Cold stress causes an *osmotic diuresis*, with sodium and chloride as the main urinary constituents. Even at a moderate level of cold stress, urine volume more than doubles. The increased urine excretion *decreases blood volume* and *lowers physical working capacity*, which is already diminished by sequestration of plasma volume in constricted peripheral vessels. Both these events *increase blood viscosity* and enhance the risk of local cold injury. Sudden immersion in cold water initiates *hyperventilation* for 1-2 minutes (risk of drowning); thereafter, ventilation rate and volume decreases to levels consistent with metabolic requirements. In severe hypothermia, carbon dioxide retention causes *respiratory and metabolic acidosis.*

Hypothermia induces progressive *depression of mental functions,* starting with apathy and ending in *lethargy* and *coma* between 30 - 28°C (86 - 82°F). Except for adrenergic peripheral vasoconstriction and shivering during cold stress, there is a gradual *decrease in nerve conduction velocity* with lowered temperature. The *signs and symptoms* of hypothermia include poor coordination, stumbling, slurred speech, irrationality and poor judgement, amnesia, hallucinations, blueness and puffiness of the skin, dilation of the pupils, decreased respiratory rate, weak and irregular pulse, and stupor. At 26 - 25°C (79 - 77°F), physical activity is impossible.

B-6

4. Freezing Cold Injury (FCI)

This sort of injury *(frostbite)* occurs when the tissue temperature falls below its freezing temperature (ca -2°C; -28°F). 90% of FCI's occur in *exposed skin* and the *extremities*, with the feet being more often affected than the upper extremities. The development of FCI is primarily determined by the ambient temperature and the duration of exposure. Other factors which increase heat loss from the skin, such as wind and *skin wetness*, increase the danger of FCI. The pathogenic mechanisms seem to involve both the direct effect of ice crystal formation *(cryogenic effect)* as well as vascular damage due to vasoconstriction-induced *vascular stasis,* leading to *tissue hypoxia*. With slow cooling, the ice crystals form in the extracellular space. This causes an increase in the osmolality of the remaining fluid which will result in water being drawn out of the intracellular compartment. The resultant *cell dehydration* changes protein structures, alters membrane lipids and cellular pH, and hence, cellular function. With fast cooling e.g., on contact with metal, they form intracellulerly, leading to death.

5. Non-Freezing Cold Injury (NFCI)

Non-freezing cold injuries (NFCI) occur when temperatures above freezing *(from 15 °C (59 °F) down to -0.5 °C (31 °F))* are maintained for long periods of time. In NFCI, the normal thermoregulatory responses of the injured body part are disturbed, and full recovery can take several years. All types of tissue may be damaged by sustained cooling, but the most serious effects occur in the nerves and blood vessels, particularly those running superficially. Sensory and autonomic fibres are the most exposed and are affected before motor coordination. At a local temperature of 7°C

(45°F), both sensory and motor nerves are blocked, resulting in *loss of sensation and paralysis*, respectively. Immersion in water is not essential to the occurrence of NFCI, but due to the high specific heat and thermal conductivity of water, NFCI is often found in combination with wet skin. NFCI often occurs in the extremities, especially the feet *(trench or immersion foot)*. However, it may occur in other areas, such as the buttocks if a subject is sitting for many hours. In general, the lower the temperature the shorter the time needed for damage to occur. Some protection against NFCI may be obtained from *cold-induced vasodilation (CIVD)* both in the hands and feet. CIVD is largely a *locally mediated response* wherein, as a result of cold paralysis, the arterioles cannot maintain tone. Blood flow and heat input are temporarily re-established until the vascular smooth muscles again contract. CIVD occurs to different degrees in people who are indigenous to cold and people who are exposed to cold for limited times. Observations suggest that cold injury is more prevalent during sojourn at high altitude than at low, for various reasons. These may include: reduced heat production as energy expenditure is diminished, higher haematocrit, raised viscosity, and decreased plasma volume.

B-7

6. Alcohol and Cold

Alcohol is a dominant cause of death in *urban hypothermia*. The ingestion of alcohol gives a feeling of warmth due to *peripheral vasodilation*. Heat loss is particularly enhanced during extended exposure to cold environmental situations in association with strenuous exercise. Acute alcohol consumption interferes with numerous metabolic processes. It *inhibits glucose-induced insulin secretion* and *stimulates pancreatic glucagon secretion*. Ethanol has in itself a calorigenic effect but requires 12% more oxygen than glucose to produce the same amount of heat. On the other hand, alcohol *reduces thermogenesis* by delaying the onset of shivering and reducing its duration. Alcohol tends to impair the efficacy of central thermoregulation. The decreased thermogenesis combined with increased heat loss despite falling central core temperature infer lowering of the set point of the thermoregulatory mechanism. Alcohol inhibits secretion of the antidiuretic hormone and, thus, *increases cold diuresis*. The impaired efficacy of thermoregulation effects both heat loss and heat production, rendering the organism poikilothermic. Not all effects of alcohol on body temperature regulation are deleterious. Thus, alcohol not only facilitates the induction of hypothermia, but enables subjects to be cooled to a lower temperature before ventricular fibrillation or cardiac arrest appears. Furthermore, ethanol facilitates defibrillation during rewarming.

B-8

7. Winter Mortality and Cardiovascular Disease

Increased mortality during the colder winter months due to *cardiovascular disease* (CVD) has been reported throughout most parts of the world, including the southern hemisphere. For example, in Norway and the Republic of Ireland, annual mortality due to CVD accounts for ca. 45% and 47%, respectively, of all registered deaths, and is about 22% and 35%, respectively, higher in December than in August (Fig. 2, left panel). There is also a clear correlation between death rate and air temperature (Fig. 2, right panel). While there are associations between (a) high CVD mortality during the winter time and (b) high CVD mortality in the coldest parts of Europe and other countries, the actual reason for the excess number of deaths in winter is unknown. It seems that most "cold exposure"-related deaths do not actually take place during exposure to cold, but occur hours or even a day or two later. This delay has been interpreted as being due to *thrombosis* starting during or shortly after cold exposure, rather than death from the immediate, reflex effects of cold.

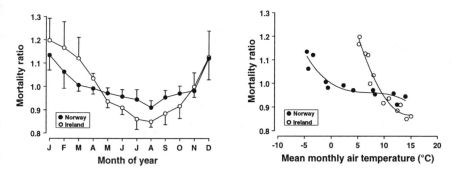

Fig. 2. Mortality from cardiovascular disease in Norway and the Republic of Ireland. Left panel: Seasonal variations. Right panel: Relation to mean monthly air temperature. Each data point represents the average monthly value for a ten year period (1985-1994) and includes data for both men & women 60 years and older.

Recent epidemiological evidence suggests that relatively brief outdoor exposure to cold conditions may be linked to increased winter mortality rate in the elderly. Acute exposure to cold in humans is well known to be associated with a variety of autonomic responses including peripheral vasoconstriction, shivering, increased heart rate and blood pressure. In patients suffering from ischaemic heart disease, acute exposure to cold may cause a decrease in coronary blood flow, often associated with coronary spasm and chest pain and even myocardial infarction. Excess winter mortality may be related to infectious diseases, such as influenza and respiratory tract infections, with an associated winter increase in the activation of the acute-phase response, including an increase in plasma fibrinogen, a well-known *risk factor*. However, several cross-correlation studies have shown that winter infections and other factors, such as seasonal variations in fat consumption and hypertension, can account for only about one third of the excess winter deaths. The weather type associated with the highest risk of heart attack appears to be cold and moist weather with low atmospheric pressure. Nevertheless, the mechanism by which seemingly mild exposure to cold ambient conditions can increase the risk of death remains unclear. Detailed analysis of the relationship between temperature changes and deaths from CVD suggests that the relative *inefficiency of autonomic control* in the elderly (see next section) permits a decrease in body temperature which, in turn, gives rise to some change which eventually leads to death. The change is probably humoral and, if it could be identified, it might be possible to prevent its occurrence or counteract its effects.

B-10

8. Autonomic Responses and Ageing

The ability to maintain body temperature, to cope with variations in ambient temperature and to develop fever is compromised with *ageing* (see also Chap. 9.2). Among the changes reported to occur with ageing are a reduction in amplitude of the diurnal body temperature oscillation *(diurnal rhythm)* as well as in the general level at which body temperature is regulated. In addition, the capability of retaining heat via peripheral vasoconstriction while awake decreases with age.

Elderly people tend to spend more time indoors and presumably rarely experience a degree of cold exposure which leads to a strong stimulation of the autonomic nervous system caused by a relatively rapid fall in body temperature (classical cold defence response to acute cold exposure with vasoconstriction and shivering). In order to explain the observation that cold exposure is in someway connected with increased winter mortality in the elderly, it should be emphasized that the way in which cold is exerting this effect may be very subtle. Here it is worth reminding ourselves of the point made in the introduction to this chapter that humans are "tropical animals" and, even at normal room temperature, it is possible to become hypothermic. Slow changes in body temperature may occur when we become inactive, for example after sitting or lying quietly for a while in a cool room. Most people will appropriately respond to such a fall in body temperature. However, the elderly often fail to detect slow rates of body cooling and, under certain circumstances, may experience a fall in body temperature to hypothermic levels without an appropriate autonomic response *(insidious (silent) hypothermia)*, due to the purported relative inefficiency of autonomic control. In this respect, housing standard, particularly with regard to heating

B-11

and insulation, as well as other socio-economic considerations, may be of importance. For example, in Fig. 2 (right panel), the difference in the relationship between cardiovascular mortality and outside air temperature in Norway and the Republic of Ireland may simply reflect the fact that housing standard is better in Norway. Thus, when we are talking about the danger of cold exposure in the elderly, we should be aware of climatic conditions inside their living quarters.

9. Summary

In this short overview, we have examined the pathophysiological state known as hypothermia as well as different types of cold injuries. Clearly in such a short chapter, it is not possible to cover all aspects in great detail and, among other things, is has not been possible to deal with the various treatment regimes both for mild and severe hypothermia as well as for treatment of the different kinds of freezing and non-freezing cold injuries. For further information, the interested reader is referred to the suggested reading list.

Acknowledgements

Fig. 2. was provided by Helene Eng, Department of Medical Physiology, Institute of Medical Biology, University of Tromsø, Norway and is based on registered data from the Norwegian and Irish Central Statistical Offices as well as the Norwegian and Irish Meteorological Offices.

10. References

1. Golden FstC. (1972) Accidental Hypothermia. *J.R. Nav. Med. Serv.* 1972, 58:196- 206:
2. Granberg, PO (1991). Human physiology under cold exposure. *Arct. Med. Res.* **50**: Suppl. 6:23-27.

11. Suggested Readings

1. *Cold Physiology and Cold Injuries* (1991). Proceedings of the Nordic Conference on Cold, Tromsø, Norway. *Arct. Med . Res.* **50**:Suppl.6.
2. *Effects of Cold on the Human Organism* (1995). Proceedings from the International Conference on Effects of Cold on the Human Organism, Oulu, Finland. *Arct. Med. Res.* **54**:Suppl.2.
3. Elwood PC, Beswick A, O'Brien JR, Renaud S. et al. (1993). Temperature and risk factors for ischaemic heart disease in the Caerphilly prospective study. *Br. Heart J.* **70**:520-523.

4. Fergusson JF, Epstein F and van de Leuv J (1983). Accidental hypothermia. *Emergency Medicine Clinics of North America* 1983; 1: 619-637.
5. Hayes P. (1989). A physiological basis of cold protection. In: Mercer JB (ed) *Thermal Physiology 1989*. Elsevier Science Publishers, Excerpta Medica, Amsterdam.pp. 45-61.
6. Lloyd EL (1986). *Hypothermia and Cold Stress*. Aspen, Maryland.
7. Reuler JB (1978). Hypothermia: pathophysiology, clinical settings and management. *Ann. Intern. Med.* **89**:519-527.
8. Davenport, J. (1992). *Animal Life at Low Temperature*. Chapman & Hall, London.
9. R. S. Pozos, P. A. Iaizzo, D. F. Danzl, and W. T. Mills, Jr. (1996). Limits of Tolerance to Hypothermia. In: Fregly MJ, Blatteis, CM (eds) *American Physiological Society Handbook of Physiology, Sec. 4: Environmental Physiology*. Oxford U. Press: New York, pp. 557-575.

Self-evaluation questions for chapter 12.2

1. Why can humans be described as tropical animals?
2. Define normothermia and hypothermia.
3. Define different types of hypothermia based on severity and duration.
4. Describe why wind chill and extremity cooling may be important factors threatening survival in the cold.
5. What are the main physiological responses to cold?
6. What is the difference between shivering and nonshivering thermogenesis?
7. Describe the main effect of hypothermia on the circulatory system.
8. What effect does cold exposure have on kidney function?
9. Compare and contrast freezing and non-freezing cold injuries.
10. Why should one avoid intake of alcohol when exposed to cold?
11. Describe how cold affects known risk factors for cardiovascular disease.
12. How does ageing affect the physiological responses to cold?

Chapter 13

Learning Objectives
1. To appreciate that the stability of T_c is maintained by the interaction of various regulatory systems that also maintain other controlled variables.
2. To recognize that the constancy of T_c is threatened when stimuli simultaneously disturb T_c and other controlled variables.
3. To know how the regulatory systems that subserve T_c and sleep affect one another.
4. To know how the thermal changes associated with the menstrual cycle influence the T_c of eumenorrheic women.
5. To know how the several environmental stressors that co-exist at high altitude differentially affect homeostatic responses.
6. To understand how diseases and drugs may disturb the stability of T_c.

Bullets

B-1 Concurrent stressors compete for shared regulatory systems.
B-2 The stability of T_c is vulnerable under certain conditions.
B-3 T_c is lower during NREMS than during wakefulness.
B-4 Warmth induces NREMS.
B-5 Sleep and fever share common, yet separable mechanisms.
B-6 The T_c of women depends on their specific menstrual cycle phase.
B-7 IL-1β may contribute to the elevated T_c of women during the luteal phase of the menstrual cycle.
B-8 T_c is under triple threat at high altitude.
B-9 Hypobaric hypoxia alters thermoregulatory responsiveness.
B-10 NST is especially susceptible to hypoxic depression.
B-11 Drugs are common causes of disordered T_c.

Chapter 13

TEMPERATURE REGULATION IN SPECIAL SITUATIONS

CLARK M. BLATTEIS
Department of Physiology and Biophysics
The University of Tennessee, Memphis 38163, USA

1. Introduction

During life, homeotherms are continually exposed to a wide variety of stimuli that challenge the constancy of their internal environment. To prevent undue deviations from homeostatic levels, corrective measures are constantly instituted. These typically involve interactions among several regulatory systems, the combined effect of which, thus, is to maintain steady-state. However, when concurrent stressors compete for shared regulatory systems, the constancy of all the affected controlled variables becomes difficult to maintain. Consequently, there occurs a hierarchy of importance, such that the constancy of some variables is sacrificed in order that that of others may be preserved.

B-1

The thermoregulatory system is particularly vulnerable to the clashing demands on different regulatory systems when two or more stressors occur concurrently. That is because, as was mentioned at the outset (see Chap. 2), apart from thermosensors and, in some species, sweat glands, it is not constituted by exclusive, individually specialized organs. Rather, the various effectors utilized to maintain T_c are components of other organ systems, *e.g.*, the cardiovascular, respiratory, muscular, and other systems. Consequently, it must compete for these effectors when stimuli occur that simultaneously perturb T_c and other controlled variables, the homeostatic defense of which requires the same effector systems. It is evident, therefore, that a potential for morbidity and even mortality exists in situations in which the competition for shared effectors can not be adequately met. By the same token, it follows that, if any one of the common effector mechanisms is impaired by, *e.g.*, disease, not only may the eventual compensatory response to a thermal challenge be altered, but so also the normal, unstimulated, cenothermic T_c. The purpose of the present chapter is to illustrate, by a few examples, how combined thermal and other stressful stimuli that impose concomitant demands on shared defense mechanisms may impact on temperature regulation. The reader wishing more details or interested in other conditions is referred to this chapter's suggested reading list and to the international Thermophysiology WWW Server at URL http://physio1.utmem.edu/THERMOPHYSIOLOGY/.

B-2

2. Some Physiological Modifiers of T_c
2.1. Endogenous
2.1.1. *Rhythms*
2.1.1.1. *Sleep-wakefulness cycle*

Thermoregulatory effector mechanisms are arousal state-dependent. Thus, thermoregulation and sleep are closely linked. Sleep is divided into two major stages, nonrapid eye movement sleep (NREMS) and rapid eye movement sleep (REMS); these states are further subdivided. During NREMS, thermoregulatory control mechanisms are intact, but the threshold T_c at which cold defense mechanisms are activated gradually decreases during the transition from wakefulness to NREMS, so that T_c is regulated at a lower level than during wakefulness (Chap. 2). During REMS, thermoregulation is suppressed, apparently because the thermosensitivity of POA neurons is greatly diminished during this state [1]. Consequently, T_c may change rapidly just after entry into REMS in a passive, poikilothermic-like fashion, the magnitude and direction of the change depending on the T_a. Both these sleep-coupled changes in T_c persist across the circadian rhythm, during fever, and in other conditions. The mechanisms underlying these changes are not known. They are self-limiting, however, because REMS episodes are normally short-lived and large shifts in T_c would trigger arousal, thus inducing corrective thermoregulatory responses.

B-3

Exposure to moderately warm T_a induces NREMS that often extends beyond the heating period. Local heating of the POA also evokes NREMS. It has, therefore, been assumed that this effect on sleep is mediated by thermosensitive neurons in this brain region. Indeed, some preoptic warm-sensitive neurons fire more frequently during NREMS. It was consequently suggested [2] that these may be GABA-ergic inhibitory neurons to the posterior hypothalamus, which is involved in the maintenance of wakefulness. Indeed, their ablation raises T_c and reduces sleep.

B-4

Preoptic thermosensitive neurons are responsive to cytokines (see Chap. 10). Therefore, it may be anticipated that cytokines would be involved in NREMS regulation as well as in fever production. Indeed, the circulating levels of TNF-α and IL-1β vary with the sleep-wake cycle, central or systemic administration of either cytokine induces NREMS, and both cytokines account for the enhanced somnogenesis associated with infections. Nevertheless, although they share common regulatory elements, sleep and fever mechanisms are distinct in that they can be separated experimentally. Thus, a variety of pyrogenic mediators (*e.g.*, PGE_2, CRH) inhibit sleep, while certain putative somnogenic mediators (*e.g.*, NO) do not induce fever. Sleep deprivation enhances the production of TNF-α and IL-1β, yet is cryogenic. IL-6 is pyrogenic, yet lacks somnogenic activity.

B-5

In sum, the relationship between thermoregulation and sleep is complex. These functions share independent regulatory systems, and their interactions result in variously altered thermoregulatory responses, depending on the sleep stage and sundry other variables.

2.1.1.2. *Ovulatory cycle*

The T_c of eumenorrheic women changes periodically according to the ovulatory cycle, being higher during the ovulatory (mid-luteal) and post-ovulatory (luteal) than during the preovulatory (follicular) stages. This is because endogenous estradiol and progesterone modulate temperature regulation. Thus, systemic progesterone increases T_c by decreasing the activity of warm-sensitive neurons in the POA. Conversely, systemic estradiol decreases T_c by increasing their firing rates. Both serum progesterone and estradiol are low in the early follicular phase of the human menstrual cycle, progesterone is low and estradiol is elevated in the late follicular phase, and both progesterone and estradiol are high in the mid-luteal and luteal phases. These differences are associated with changes in thermoregulatory effectors functions, such that a downward shift occurs in the T_c threshold for sweating onset during the transition from early to late follicular phase, while an upward shift occurs in the T_c thresholds for both heat- and cold-defense mechanisms during the change from late follicular to mid-luteal phase. Thus, the onset of thermoeffector responses to heat, cold, or exercise of women, and their regulated T_c during such thermal changes are affected by the reproductive hormonal profile associated with their specific menstrual cycle phase. Concomitant, menstrual cycle-related shifts in plasma volume dynamics, vasomotor, and other homeostatic control systems also contribute to these thermoregulatory differences.

B-6

It is of interest to note in this context that, while the above changes in thermoregulatory functions during the menstrual cycle are consistent with the central mechanism of action by reproductive steroids described earlier, it is not clear whether these hormones can cross the BBB and act directly on thermosensitive neurons in the POA. Although progesterone and estrogen receptors have been characterized and mapped within the (rat) brain, it is possible that these factors nevertheless act indirectly via secondary mediators. Indeed, progesterone and estrogen have been shown to stimulate IL-1 and PGE_2 productions at low doses and to inhibit them at high doses, and the plasma activity of IL-1β increases and that of IL-1ra decreases during the luteal phase of the menstrual cycle, when T_c is regulated at a higher level. Thus, the changes in thermoregulatory effector functions during the ovulatory and post-ovulatory phases of the menstrual cycle could be due to the direct action of the reproductive hormones in the POA or to the indirect action of secondary mediators, such as IL-1.

B-7

In any case, these findings serve to reinforce the notion that thermoregulatory adjustments are modulated by multiple, concurrent interacting drives.

2.2. Exogenous
2.2.1. *Environmental*
2.2.1.1. *Terrestrial altitude*

Three stressors normally co-exist in the natural environment of terrestrial altitude: 1) decreased barometric pressure (P_b; *hypobaria*), which by consequence brings on 2) a lowered partial pressure of ambient oxygen (P_{O_2}; *hypoxia*), and 3) a T_a that is progressively reduced (*cold*) with elevation above sea level. Wind at high speeds may be an aggravating factor of the last element; the relative humidity, however, is usually low, but solar radiation is augmented. Each of these environmental stressors causes *per se* profound disturbances to the steady-state of several controlled variables, each of which utilizes common regulatory systems for corrective action. Thus, major readjustments of the cardiac output to various tissues, mediated by differentially distributed sympathetic discharges and local autoregulatory factors, and hyperventilation and the consequent hypocapnemia and respiratory alkalosis represent other compensatory responses to hypoxia that affect the stability of T_c at high altitude. It may be anticipated, therefore, that the efficiency of the thermoregulatory system will depend partly on the immediate importance for survival of the various functions being defended and partly on the relative strength of each of the affecting stimuli, creating potential conflicts, with possibly deleterious effects.

From the perspective of thermal homeostasis only, the hypobaria of altitude directly alters the insulative value of the thermal boundary layer (I_a; see Chap. 3 and 5). Thus, on the one hand, the lowered density of molecules at the interface between skin surface and ambient air reduces the latter's capacity to convect heat away from the body, *i.e.*, it increases the value of I_a, particularly if the wind speed is low; in consequence, mean T_{sk} at rest in thermoneutrality is higher at altitude than at sea level. This result is beneficial when passively exposed in a cold environment, but less desirable if working in it. Mitigating this effect, on the other hand, evaporative heat loss is enhanced at high altitude due to the augmented water vapor pressure difference between wet (sweating) skin and the combined hypobaria and cold-associated low relative humidity, an advantage when climbing and exposed to the more intense solar radiation at high terrestrial altitudes. Consequently, T_c does not rise excessively under such conditions. However, the thus increased water loss does entail a risk of hypohydration, a further stress requiring corrective actions.

The hypoxia of altitude also affects the body's thermal homeostasis. It may be expected that if the O_2 supply to thermogenic tissues (e.g., skeletal muscle, BAT; see Chap. 4.1 and 4.2) were reduced, the ability to provide sufficient heat to maintain T_c would be similarly reduced. Indeed, when the ambient P_{O_2} declines below a critical level (ca. 60 torr or <8% inspired O_2 at sea level), the resting metabolic rate at neutral T_a is depressed proportionally to the degree of hypoxia; as a result, T_c may fall despite the reduced convective heat loss at altitude. This hypoxic reduction becomes manifest at higher threshold values of P_{O_2}, however, if the metabolic rate is above its resting level; i.e., the increase in metabolic rate normally evoked by cold exposure is particularly vulnerable to reduction by hypoxia, the more so the higher the metabolic rate. Because shivering continues largely unimpaired, and may even be augmented under these conditions, and because the increase in metabolic rate produced in room air at thermoneutrality by infusions of norepinephrine, the mediator of nonshivering thermogenesis (NST; see Chap. 4.2), is abolished by moderate hypoxia (10-12% O_2), it would seem that hypoxia depresses thermoregulatory heat production mainly by selectively inhibiting NST, probably by reducing the oxidative capacity of, particularly, BAT. Neonates and smaller animals, consequently, are especially susceptible to this effect. Although adult humans and other large mammals possess little BAT, regulatory NST by other organs nevertheless contributes to an appreciable extent (15-20%) to thermogenic processes, so that, in fact, their cold-induced increase in metabolic rate is also reduced at altitude as compared to sea level. Interestingly, this altitude-induced reduction of the thermogenic response to cold is not reversed by prolonged residence at altitude, i.e., by altitude acclimatization, but is immediately reversible on descent to sea level. Interestingly also, most hypoxic animals select behaviorally environments that are colder than do normoxic ones, thereby further facilitating the decrease in their T_c. To the extent that this hypoxia-induced hypothermia may lower the metabolic rate and, thus, the need for O_2 when it is in short supply, this behavior may be considered beneficial.

B-10

Hypoxia-induced changes in cutaneous circulation also modify thermal homeostasis at thermoneutrality. Vasoconstriction of the extremities occurs on acute exposure to hypoxia, but is superseded by vasodilation as residence continues. Skin blood flow of the proximal limbs and trunk, however, is elevated from the outset. The large surface area of this portion of the body accounts for the high mean T_{sk} at altitude. The vasomotor responses to cold and heat apparently proceed normally, albeit more sluggishly.

From the above, it may be anticipated that the hypoxic depression of metabolic rate above the basal would not be limited to cold-induced thermogenesis, but would similarly affect that due to other thermogenic processes. Indeed, hypoxia also reduces pyrogen-induced thermogenesis, hence attenuating fever. The usual cutaneous vasoconstriction that accompanies fever production (see Chap. 10), moreover, is reduced by hypoxia.

2.2.1.2 Other environmental stressors

It may be surmised from the preceding that exposure to other special environments may similarly induce systemic physiological changes that impact on thermal homeostasis. For example, exposure to hyperbaric and aquatic environments affect heat exchange mechanisms. Compensatory responses may be further modified by the particular gas mixture breathed under these conditions, the individual heat-conductive properties of which (*e.g.*, helium) are independent of P_b. Likewise, other combinations of disparate stressors, *e.g.*, heat and exercise (see Chap. 8), heat, fever and dehydration, etc., may be expected to impact synergistically on competing regulatory mechanisms such that thermoregulation may become compromised. Space limitations do not permit further consideration of such situations here. The interested reader is referred to the suggested reading list at the end of this chapter.

3. Some Pathophysiological Modifiers of T_c

3.1. Endogenous

3.1.1. *Disease*

Malfunctions that affect the body's ability to conserve or to produce heat may be expected to cause an imbalance in body heat storage, hence in T_c. Some illnesses that thus compromise thermoregulatory mechanisms at different levels of their organization are listed in Table 1. Please consult the suggested reading list for more details about the alterations involved. Diseases associated with fever are not included in this context.

Table 1. SOME MEDICAL CONDITIONS THAT AFFECT THE STABILITY OF T_C

A.   **Modify heat exchange mechanisms**
       1.   *Increase heat loss*
             Skin disorders (certain lesions; erythrodermas; extensive burns or wounds; cutaneous manifestations of certain systemic diseases); peripheral vascular disease (Raynaud's; a-v shunts; phlebitis; neuropathies; heavy smoking); cardiopulmonary disease.
       2.   *Decrease heat loss*
             Skin disorders (certain lesions; epidermal thickening; extensive burn scars; cutaneous manifestation of certain systemic diseases); peripheral vascular disease (atherosclerosis; microangiopathies); sweat gland dysfunctions; diabetes insipidus; obesity; deconditioning (bedrest; prolonged confinement).

B.   **Modify thermogenic mechanisms**
       1.   *Increase heat production*

                Hypermetabolic disorders, secondary to endocrinopathies (thyrotoxicosis; granulomatous thyroiditis; pheochromocytoma; 1° and 2° adrenal insufficiency); seizure activity; agitated or tremulous states.

    2.   *Decrease heat production*

            Paralysis; paraplegia; incapacitating trauma; debilitating illness; neuromusucular and joint disorders; dystonia; depleted energy stores (malnutrition; anorexia nervosa; sleep deprivation; cachexia); hypometabolic disorders, secondary to endocrinopathies (hypothyroidism; hypopituitarism; diabetes mellitus; hypoglycemia); degenerative diseases of skeletal muscle; circulatory shock of various etiologies (hemorrhagic; septic; etc.); secondary to certain systemic diseases (uremia; pancreatitis; systemic acidosis; hepatic failure; renal impairments; obstructive airway disease).

C.   **Modify thermoregulatory control**
    (May increase or decrease T_c)

            Brain and spinal cord lesions (traumatic; vascular; neurochemical; infectious; neoplastic); neurologic disorders (autonomic dysfunctions; sensory, central, or afferent pathway impairments); psychiatric illnesses; stress; encephalopathies of different etiologies.

3.2. Exogenous

3.2.1. *Drugs*

In view of their widespread actions on physiological processes generally, it is little wonder that drugs affect the ability to regulate T_c. Drug effects are complex, not well understood, and influenced by multiple, both internal (*e.g.*, body mass, level of activity, individual sensitivity) and external (*e.g.*, dose, route of administration, T_a) variables. As a detailed consideration of this topic is beyond the scope of this text, interested readers are referred to the reviews cited in the suggested reading list. Some representative drugs are listed in Table 2 by their reported actions on thermal homeostasis. Among these drugs, attention is called to the disrupting effects of drugs of abuse, *e.g.*, ethanol, opioids. Note also that several of the drugs may have opposite actions, depending on various factors (not listed here) as well as impair T_c stability by acting at several levels of regulation simultaneously, *e.g.*, tricyclic antidepressant, opioids. Finally, some drug categories (*e.g.*, certain muscle relaxants, volatile anesthetics) may trigger very rapid T_c rises in susceptible subjects (*e.g.*, malignant hyperthermia). Not included in this brief review are complications related to pyrogenic contaminants or to the administration of potentially irritating drugs (*e.g.*,

Table 2. SOME CLASSES OF DRUGS THAT AFFECT THE STABILITY OF T_C

Drugs	Heat Loss +	Heat Loss −	Thermogenesis +	Thermogenesis −	CNS +	CNS −	Tc +	Tc −
General anesthetics	√			√		√		√
Antinicotinics (e.g., tubocurarine)				√				√
Antimuscarinics (e.g., atropine)	√	√						√
Adrenergic blockers (e.g., phentolamine)	√			√		√		√
Adrenergic agonists (e.g., cocaine, amphetamines, MOA inhibitors)		√	√		√		√	
Benzodiazepines (e.g., diazepam)	√			√		√		√
Barbiturates (e.g., phenobarbital)	√			√		√		√
Phenothiazines (e.g., chlorpromazine)	√			√	√	√	√	√
Opioid peptides (e.g., β-endorphin)	√	√	√		√	√	√	√
Butyrophenomes (e.g., haloperidol)	√					√		√
Chemical toxicants (e.g., ozone)	√			√	√			√
Ethanol	√			√		√		√
Tricyclic antidepressants (e.g., imipramine)	√	√	√		√	√	√	√
Antihistamines (e.g., cyproheptadine)		√						√
Hallucinogens (e.g., LSD)		√	√		√		√	
Antiparkinsonians (e.g., amantadine)	√				√		√	

+: increase; −: decrease

erythromycin) or cytoxic (*e.g.*, bleomycin) and other agents that may induce endogenous pyrogens.

4. Therapeutic Hyperthermia and Hypothermia

The application of exogenous heat or cold to raise or lower tissue temperatures, respectively, as adjunctive measures in the treatment of certain diseases has come into increasing use in recent years. They are briefly mentioned here, for completeness.

4.1. Heat

Heat is used in cancer therapy. The biological basis for its use relates to the greater hyperthermic cytotoxicity of malignant than of normal cells. Hyperthermia also enhances the therapeutic effects of radiotherapy and chemotherapy. Clinical hyperthermia is induced by nonionizing electromagnetic waves or ultrasound, non-invasively (*e.g.*, superficial microwave applicators), minimally invasively (*e.g.*, intracavitary devices), or very invasively (*e.g.*, interstitial implants).

4.2. Cold

Regional or total body hypothermia (15-20 °C) is used most commonly in surgical applications. The rationale for its use is based on the hypothermic decrease in metabolic rate, which protects tissues and organs from hypoxic injury due to the interruption of blood flow during their repair or transplantation (*e.g.*, kidney). Extreme hypothermia (cryosurgery) is also used to scar or destroy tissue (*e.g.*, retina). Cooling (and rewarming) is accomplished by means of wrapping the body in a circulating (water or air) blanket, or by directly varying the blood temperature via heat exchangers attached to the pump oxygenator. Mild hypothermia (a 2-6 °C reduction of T_c) may also be of benefit in reducing pain (cryoanalgesia). It also may afford protection against the toxic and possibly lethal effects of xenobiotic agents, as well as increase survival following ischemia- or trauma-related brain injury.

5. Summary

The near constancy of T_c is maintained by a complex interaction of autonomic, humoral, and behavioral mechanisms involving the cardiovascular, respiratory, neuromuscular, endocrine, and other regulatory systems. Stimuli that disturb controlled variables regulated by any of these systems may be expected, therefore, to impinge as well on the stability of T_c. When such stimuli simultaneously disturb T_c and other controlled variables, the homeostatic defense of T_c is accommodated to the competing demands of the combined stressors for shared regulatory systems. Under such conditions, the regulation of T_c occupies a rank in a hierarchy of homeostatic controls, determined partly on the importance for survival of the functions being defended and partly also on the relative intensities of the disturbances; a rank, therefore, that may change depending on the coexistent conditions. Examples of some representative conditions that thus affect thermal homeostasis were briefly considered in this chapter.

Suggested reading:
Blatteis CM (ed) (1997). Thermoregulation. *Ann. NY Acad. Sci.* 813: 1-878.

Blatteis CM, Lutherer LO (1976). Effects of altitude exposure on thermoregulatory response of man to cold. *J. Appl. Physiol.* 41: 848-858.

Blatteis CM, Ungar AL, Howell RB (1994). Thermoregulatory responses of rabbits to combined heat exposure, pyrogen and dehydration. In: Milton AS (ed), *Temperature regulation.* Birkhäuser, Basel, pp. 71-74.

Doherty Jr DW, Blatteis CM (1980). Hypoxic reduction of endotoxic fever in guinea pigs. *J. Appl. Physiol.* 49: 294-299.

Fregly MJ, Blatteis CM (eds) (1996). *Handbook of physiology, sec. 4: Environmental physiology, vols. I and II.* Oxford Univ. Press, New York.

Gordon CJ, Revani AH (1995). The role of temperature in neurotoxicity. In: Chang LW, Dyer RS (eds), *Handbook of neurotoxicity.* Marcel Dekker, New York, pp. 1049-1068.

Grecza R, Smorawinski J (1989). Thermoregulatory response to exercise in women before and after ovulation. In: Mercer JB (ed), *Thermal physiology 1989.* Elsevier Science, Amsterdam, pp. 341-345.

Heller HC, Glotzbach SF (1977). Thermoregulation during sleep and hibernation. *Int. Rev. Physiol.* 15: 147-187.

Hessemer V, Brück K (1985). Influence of menstrual cycle on thermoregulatory, metabolic and heart rate responses to exercise at night. *J. Appl. Physiol.* 59: 1911-1917.

Hessemer V, Brück K (1985). Influence of menstrual cycle on shivering, skin blood flow, and sweating responses measures at night. *J. Appl. Physiol.* 59: 1902-1910.

Krueger, JM, Majde JA (1994). Microbial products and cytokines in sleep and fever regulation. *Crit. Rev. Immunol.* 14: 355-379.

Mackowiak PA (ed) (1987). *Fever*: *Basic mechanisms and management*, 2nd ed. Lippincott-Raven, Philadelphia

McGinty DJ, Szymusiak RS (1990). Hypothalamic thermoregulatory control of slow-wave sleep. In: Mancia M, Marini G (eds), *The Diencephalan and sleep*. Raven, New York, pp. 97-110.

Obal Jr F (1984). Thermoregulation and sleep. *Exp. Brain Res. Suppl.* 8: 157-172.

Parmeggiani PL (1980). Sleep and temperature regulation. In: Szelényi Z, Székely M (eds), *Contributions to thermal physiology*. Pergamon, Oxford, pp. 207-215.

Schönbaum E, Lomax P (eds) (1991). *Thermoregulation. Pathology, pharmacology, and therapy*. Pergamon, Oxford.

Sessler DI (1994). Thermoregulation and heat balance: general anesthesia. In: Zeisberger E, Schönbaum E, Lomax P (eds), *Thermal balance in health and disease*. Birkhäuser, Basel, pp. 251-265.

Sessler DI (1997). Perioperative thermoregulation and heat balance. *Ann. NY Acad. Sci.* 813: 757-777.

Stephenson LA, Kolka MA (1985). Menstrual cycle phase and time of day alter reference signal controlling arm blood flow and sweating. *Am. J. Physiol.* 249: R186-R191.

Self-study questions
1. Under what conditions is the constancy of cenothermic T_c particularly threatened?
2. What factors account for the vulnerability of T_c under those conditions?
3. How well is T_c regulated during NREMS?
4. How well is T_c regulated during REMS?
5. What role may some POA thermosensitive neurons play during NREMS?
6. What are the relations between fever and sleep?
7. How do estradiol and progesterone influence the T_c of eumenorrheic women?
8. How might IL-β be involved in the regulation of T_c during the luteal phase of such women?
9. What factors account for the risk to the stability of T_c at high altitude?
10. In what ways does the hypobaria of altitude affect thermoregulatory ability?
11. In what ways does the hypoxia of altitude affect thermoregulatory ability?
12. What evidence supports the view that NST is particularly susceptible to hypoxic reduction?
13. Does hypoxia affect skin blood flow?
14. Describe how T_c may be altered by disease. Give examples.
15. List ways by which drugs may alter T_c. Give examples.

Appendix 1 (Chap. 2)
Methods of Body Temperature Measurement
Clark M. Blatteis, Department of Physiology and Biophysics, The Univesity of Tennessee, Memphis 38163, USA

1. Devices
Any device used to measure temperature is termed a *thermometer*. The functional requirements of its operation are only that it possess a sensor having some property that changes with temperature, and that this change can be visualized and calibrated against a temperature standard. In the USA, standard specifications for accuracy for clinical thermometers are periodically established by the American Society for Testing and Materials (ASTM). Some commonly used, commercially available thermometers are briefly described below. Further information can be obtained by consulting the references cited.

1.1 *Liquid-in-glass thermometers*
Various devices suitable for clinical measurements have been developed since the introduction of the first device, the air thermoscope [1]. The sealed, mercury-in-glass thermometer, developed by Fahrenheit in *ca.* 1713, is the oldest of these devices still in use. Its modern version has a small constriction below its scale which prevents the mercury from flowing back into the bulb as the thermometer cools, thus recording the *maximum* temperature reached. If correctly calibrated, it is suitably accurate for *static* measurements. The principal sources of error arise from the mercury not being shaken back below the constriction before use, thus yielding erroneously high readings, and failure to maintain the thermometer in place sufficiently long (*ca.* 3 min), thus yielding low readings (the relatively large thermal mass of the mercury and glass bulb, on the one hand, and the low thermal conductivity of glass, on the other, lengthen the time constant of the instrument). Moreover, because the thermometer has to be removed, its mercury column shaken down, and then the probe reinserted, it is unsuitable for monitoring continuous variations in T_c.

1.2 *Electronic thermometers*
Mercury thermometers are now rapidly being replaced by electronic thermometers. One popular type consists of a sensor that is, in effect, a temperature-sensitive resistor, or *thermistor*. Another is a differential thermometer, or *thermocouple*, constructed by placing one junction of two metals (typically copper and constantan) at the measuring point and the other in an ice-point cell or temperature-controlled block. The difference in temperature between the sensor in the medium being measured and that at the reference temperature generates a voltage proportional to that difference. Various forms of such devices exist [2, 3]. Their calibration variability is ± 0.1°C or better. They connect to a digital display, which minimizes error. They also

have short equalization times (*ca.* 1 min) and, thus, can record T_c continuously. A significant advantage is that they are flexible, and the probe tips can be fashioned to suit different sites on the body.

1.3 *Chemical thermometers*
Cholesteric liquid crystals are ordered in a molecular arrangement of layers such that light traces a pattern as it passes through the spaces between the layers. This molecular arrangement is strongly temperature-dependent, *i.e.*, the layers wind and unwind with temperature, and their altered spacing is reflected as a color change. The crystals exist as sprays, disks and bands, and are easy to apply, but their accuracy and reliability are low [4]. They have some clinical applications, particularly as forehead thermometers.

1.4 *Radiation thermometers*
These are noncontact thermometers that measure the radiated electromagnetic power in the infrared band (see Chap. 3). The detected signals are displayed as temperatures. Various models exist commercially. They are used chiefly to measure tympanic temperature (see later). Data can be obtained in a few seconds. They are, therefore, increasingly popular in clinical settings. However, training and experience are needed to achieve accurate results [5].

1.5 *Other thermometers*
Current research and development in thermometry involve, in particular, the adaptation of various imaging techniques to measure T_c noninvasively. These include magnetic resonance imaging signals, computed tomography Hounsfeld numbers, microwave radiometry, electrical impedance thermography, acoustic radiometry, and others [6]. Novel computerized data acquisition systems, multiple-sensor thermocouple probes, fiberoptic thermometers, and improved radio-telemetered, thermistor-based systems are also being developed.

2. Techniques
Thermometeric measurements are influenced by a variety of technical factors, *e.g.*, the type of device being used, its accurate calibration, its proper placement at the anatomic site being measured (there are regional differences related, *e.g.*, to location in the oral cavity, depth in the rectum), etc. In clinical settings, if practicable, T_c should be recorded at rest in thermoneutrality.

2.1 *Assessment of the thermal core*

2.1.1 Axillary temperature (T_{ax})

Historically, T_{ax} is the oldest technique of T_c measurement. It was used by Wunderlich, who published the first major treatise on clinical thermometry in 1868. He analyzed the T_{ax} of a large sample of healthy adults and found that it ranged overall from 36.2 °C (97.2 °F) to 37.5 °C (99.5 °F). He considered the mean to be 37 °C (98.6 °C), and his observations thus established the notion that 37 °C is the physiologically "normal" T_c. More recent studies, however, have shown that, in fact, T_{ax} is an inadequate index of the T_c because it can be affected markedly by the local blood flow, underarm sweating (both thermal and nonthermal), inappropriate positioning of the probe and/or inadequate closure of the axillary cavity (hence, by T_a), and by insufficient duration of recording (with a mercury thermometer, 5-10 min under the tightly closed axilla). However, T_{ax} may provide a closer approximation of T_c in neonates, particularly when mercury-in-glass thermometers are used [7].

2.1.2 Rectal temperature (T_{re})

T_{re} measurements were advocated by Liebermeister (1875) as being more reliable than T_{ax}. They range overall from 37.3 °C (99.2 °F) to 37.6 °C (99.6 °F), *i.e.*, higher than T_{ax}. T_{re} is widely used for assessing T_c in clinical practice, although there is a risk of rupturing the rectal or colonic wall if the probe is inserted carelessly, especially in infants. The method has also several drawbacks. Thus, because there is a gradient of temperature through the rectum, the depth of probe insertion is critical and should be standardized; in the adult human, that depth is recommended to be 4 cm. Inaccurate values may be obtained, moreover, due to imbedding in a fecalith or dislodgement during defecation. Also, because of the passage close to the posterior rectal wall of blood returning from the legs, T_{re} may be higher (as after running) or lower (as during cold exposure) than the actual T_c. Finally, the response of T_{re} to thermal stimuli, internal or external, that provoke an active imbalance between heat production and loss is delayed by comparison to other deep body temperatures (Fig. 1); this is due, in part, to the vascular perfusion of the rectum lagging that of the deeper core. Hence, T_{re} should be considered an acceptable index of T_c only in steady-state conditions, when body temperatures are not changing rapidly.

2.1.3 Oral temperature (T_{or})

T_{or} has long been the preferred measure of T_c in clinical settings in the USA. The thermometer is inserted under the tongue toward the back of the oral

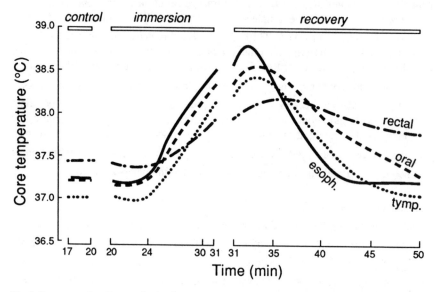

Fig. 1. T_c measured orally, rectally, in the deep esophagus, and against the tympanum of a subject acutely immersed in a waterbath at 41 °C. Note the prompt response of T_{es} on and after immersion. (From [10].)

cavity, and the probe tip positioned in a pocket at the base of the tongue, on either side of the mouth; the mouth is then closed tightly. Although the method is straightforward, there are pitfalls, *e.g.*, salivation, previous intake of hot or cold food and fluids, and smoking may affect the measurements. Also, breathing may be impeded in subjects with blocked nasal passages. The technique is also not always practicable in small children. However, T_{or} generally follows changes in T_c closely (Fig. 1), because the sublingual pockets are perfused by a branch of the external carotid artery; the temperature of blood in a large, proximal artery is analogous to the mean T_c ([see below]). T_{or} is on average 0.6 °C lower than T_{re}.

2.1.4 Esophageal temperature (T_{es})

Because the pulmonary artery conveys blood returning to the heart from both the heat-producing core and the heat-losing shell, it is generally considered that the temperature of this mixed blood approximates most closely the mean temperature of the whole body. It is also thought to be reflected very closely in the temperature of the blood flowing in the other major arteries. Unfortunately, measurement of this temperature is not practicable without invasive catheterization. The temperature of the deep esophagus has, therefore, been used as a less invasive alternative, as it closely parallels that of the pulmonary arterial and aortic blood. The measurement of T_{es} is thus the preferred method when dynamic changes in T_c need to be followed reliably (Fig. 1), e.g., in malignant hyperthermia. A lubricated, flexible thermometric probe is inserted through a nostril and passed until the tip is located behind the middle of the lower third of the sternum, at the cardia. Positioning is critical for accurate measurements because of longitudinal variations in T_{es}. A drawback is that awake subjects generally find the procedure unpleasant.

2.1.5 Tympanic temperature (T_{ty})

Although it is by no means established [8], it is nevertheless often stated that the most critical T_c, in terms of thermoregulation, is that of the brain. Since, in man, the tympanic membrane is supplied by branches of the external carotid artery, T_{ty} has come to be considered the ideal site for measuring T_c (Fig. 1). Indeed, the recent adaptation of infrared radiometers to thermometric measurements (see earlier) has caused T_{ty} lately to become the measurement of choice in routine clinical practice in the USA. The sensor is carefully inserted into the aural canal until resistance is felt, and then secured firmly so that it will remain in place; models exist commercially with features for holding the probe against the tympanic membrane with a constant force. However, mechanical pressure on the tympanic membrane is often painful, and there is a risk of perforating it. Moreover, incorrect positioning may place the sensor in the aural canal, where the temperature would be affected by heat gain from or heat loss to the outer ear, or by others thermal exchanges of a local nature. T_{ty} in routine clinical practice is, therefore, more an index of the relative than of the absolute T_c, but adequate for that purpose.

2.1.6 Other internal body temperatures

By inserting suitable temperature sensing elements into the desired site, the temperature of any locus within the body, e.g., subcutaneous, intramuscular, intravascular, abdominal organs, brain, etc., can be measured, both acutely and chronically. However, the placement methods are unavoidably invasive, thereby entailing local injury to tissues and a high risk of inflammation, then fibrosis. These effects can alter local conditions in multiple ways as well as affect the sensitivity of the sensors themselves. For specific information on procedures, the interested reader is invited to query the Thermophysiology Worldwide Web Server at the following URL: http://physio1.utmem.edu/THERMOPHYSIOLOGY/

2.2 Assement of the thermal shell
2.2.1 Skin temperatures (T_{sk})

A variety of sensing devices for measuring surface temperatures have been developed. They are available commercially in many shapes and forms. After cleaning the skin to remove dirt and grease, the sensing surface of the probe (the other side is usually insulated) is placed in contact with the skin and firmly secured in place. Small errors in T_{sk} measurements may occur, however, due to interference by the probe with heat exchange by conduction, convection, and evaporation (see Chap. 3) in the region of skin to which it is attached. Normally, several different locations on the skin are recorded simultaneously. By using weighted coefficients (based on skin surface area), the mean T_{sk} can be calculated. Various systems for measuring mean T_{sk} (location of sites, weighting factors, formulas, etc.) have been devised [9]. Similarly, by appropriately weighting (depending on the T_a) the mean T_{sk} and the T_c, the mean temperature of the whole body can be estimated [9].

3. Summary

The particular site of measurement and possible shortcomings of the technique and thermometric device being used should always be considered in addition to the dependent physiological variables when interpreting results. By the same token, the choice of a method of T_c measurement should be governed by the purpose of that measurement. Both T_{es} and T_{ty}, when the measuring probes are properly positioned, are close approximations of T_c under most conditions, while T_{re} and T_{or} are acceptably representative measures in steady-state conditions.

4. References

[1] Middleton WEK (1996). *A history of the thermometer and its use in meteorology*. The Johns Hopkins Press, Baltimore.

[2] Siwek WR, Sapoff M, Goldberg A, Johnson AC, Botting M, Lonsdorf R, Weber S (1992). Thermistors for biomedical thermometry applications. In: Schooley JF (ed), *Temperature: Its measurement and control in science and industry, vol. 6.* American Institute of Physics, New York, pp. 475-480.

[3] Cetas TC (1990). Thermometry. In: Field SB, Hand JW (eds), *An introduction to the practical aspects of clinical hyperthermia.* Taylor and Francis, London, pp. 423-477.

[4] Lees DE, Schuette W, Bull JM, Whang-Peng J, Atkinson ER, MacNamara TE (1978). An evaluation of liquid-crystal thermometry as a screening device for intraoperative hyperthermia. *Anesth. Analg.* 57: 699-704.

[5] Fraden J (1992). Medical infrared thermometry; review of modern techniques. In: Schooley JF (ed), *Temperature: Its measurement and control in science and industry, vol. 6.* American Institute of Physics, New York, pp. 825-830.

[6] Cetas TC (1997). Thermometers. In: Mackowiak PA (ed), *Fever: Basic mechanisms and management, 2nd ed.* Lippincott-Raven, Philadelphia, pp. 11-26.

[7] Schiffman RF (1982). Temperature monitoring in the neonate. A comparison of axillary and rectal temperatures. *Nurs. Res.* 31: 274-277.

[8] Jessen C (1990). Thermal afferents in the control of body temperature. In: Shönbaum E. Lomax P (eds), *Thermoregulation: Physiology and Biochemistry.* Pergamon, New York, pp. 153-183.

[9] Gonzales RR, Gagge AP (1996). Mechanisms of heat exchange: Biophysics and physiology. In: Fregly MJ, Blatteis CM (eds), *Handbook of Physiology. Sec. 4: Environmental Physiology, vol. 1.* Oxford Univ. Press, New York, pp. 45-84.

[10] Edwards RJ, Belyavin AJ, Harrison MH (1978). Core temperature measurement in man. *Aviat. Space Environ. Med.* 49: 1289-1294.

Appendix 2 (Chap. 3)
List of Symbols

Table B: List of Symbols

Symbol	Description	Units
A	area, cross section	[m^2]
a	integrated central signal about thermal state of body	[°C]
a (suffix)	ambient	
BF'	volumetric blood flow (perfusion)	[$m^3_{bl}\ s^{-1}\ m^{-3}$]
b (suffix)	body	
bl (suffix)	blood	
C	convective heat exchange with the environment	[W]
c	specific heat	[$Ws\ kg^{-1}\ °C^{-1}$]
c (suffix)	convective	
c_1, c_2	constants	
co (suffix)	core	
E	heat loss via evaporation	[W]
EFF	general symbol for an effector mechanism such as shivering, sweat rate, vasomotor changes	
e (suffix)	evaporative	
e_1, e_2	constants	
ev (suffix)	evaporated	

G	general symbol for controller gain (transfers thermal load error [°C] into effector change: shivering, sweat rate, vasomotor change)	
g	weighting factor for intracorporal temperature signals	
H	heat flow	[W]
H'	volumetric heat flow	[Wm^{-3}]
HL	heat loss	[W]
H$_k$	conductive heat flow through tissue	[W]
h	heat transfer coefficient	[W m^{-2} °C^{-1}]
I	thermal insulation	[m^2 °C W^{-1}]
i (suffix)	inner	
K	conductive heat exchange with the environment	[W]
k	conductance	[Wm^{-2} °C^{-1}]
k*	absolute thermal conductance	[W °C^{-1}]
k (suffix)	conductive	
LHV	latent heat of vaporization	[Ws m^{-3}]
M	metabolic rate	[W]
MEX	metabolic rate due to exercise	[W]
MSH	metabolic rate due to shivering	[W]
m	body mass	[kg]

o (suffix)	outer; operative; basal	
p	water vapour pressure	[mm Hg] or [Pa]
R	radiative heat exchange with environment	[W]
R	thermal resistance	[m² °C W⁻¹]
R (suffix)	respiratory	
R*	absolute thermal resistance	[°C W⁻¹]
RH	relative humidity	[%]
r (suffix)	radiant, radiative	
S	heat storage	[W]
SWR	sweat rate	[m³ s⁻¹]
sat (suffix)	saturated	
sk (suffix)	skin	
T	temperature	[°C]
tot (suffix)	total	
v	wind speed	[m s⁻¹]
VM	vasomotor changes	
$\dot{V}_{O_2}$	rate of O₂ consumption	[ml kg⁻¹ min⁻¹]
W	mechanical power produced by exercise	[W]
wet (suffix)	wetted	
x	local coordinate	[m]

α	control factor	
Δ...	deviation of ..., difference of ...	
ε	emissivity	
λ	thermal conductivity	[W m^{-1} °C^{-1}]
ρ	density	[kg m^{-3}]
σ	Stefan-Boltzmann constant	[W s^{-1} K^{-4}]

INDEX

"J" wave 250
5,6 dihydroxytryptamine (5,6-DHT) 222
5-HT 222
6-hydroxydopamine (6-OHDA) 222
α1-receptors 87
accidental hypothermia 246
acclimation 194, 195, 208
acclimation to cold 73, 133
acclimatization 169, 194, 208
acetylcholine 56, 165
active vasodilation 87
acute cold 74
acute hypothermia 246
acute-phase reaction 178
acute-phase response 254
adaptation 95, 117, 194, 203, 204
adaptive hypothermia 216, 223
adipocytes 66, 67
adrenal hormones 213
adrenergic receptors 84
adrenergic responsiveness 200
adult humans 73, 74, 75
aerobic capacity 130, 136
afferent signaling 181
age 130, 134, 138
ageing 119, 139, 254
air 86
Alacaluf Indians 210
alcohol 115, 116, 167, 168, 239, 246, 252
Allen's rule 211
alliesthesia 112, 113, 118
α-melanocyte stimulating hormone 155
altitude 187, 263, 264
altitude acclimatization 264
altricial 71
Alzheimer's disease 166
ambient temperature 15, 80
amyotrophy 118
anapyrexia 115, 121

anastomoses 87
anesthesia 117, 222
anhidrosis 119
anorexia nervosa 122
antidiuretic hormone 102
antipyretic 188
antipyretic drugs 8
antipyretics 184
antipyretics 9
APR 187, 188
arachidonic acid 183
arginine vasopressin 155
arousal 217
arteriovenous 87
arteriovenous anastomoses (AVAs) 28, 88, 212, 235, 247
arterio-venous heat exchange 28
arthritis 162, 167
ATP 68
atrial fibrillation 250
auditory canal temperature 136
autonomic 95, 110, 111
axillary temperature 275

β1-adrenergic receptors 71, 67
β3-adrenergically-mediated 71
β3-receptors 67
β3-subtype 76
Barbour 5
basal metabolic rate (BMR) 210
brown adipose tissue (BAT) 149, 150, 151, 214, 215, 223
BBB 262
bed rest 136
behavior 80, 85, 108, 109, 110, 111, 114, 118, 223, 224, 264
behavioral 86, 95, 167, 208
behavioral self-adjustment 110
behaviors 101, 208

behavioural mechanisms 63, 153
behavioural responses 152
benefits and risks 187
benzodiazepines 117
Bergmann's rule 211
Bernard 6
biofeedback 116
biophysics 25
birds 64, 204
birth 147, 149, 152, 155, 157
black body 31
blood 132
blood composition 250
blood flow 132, 133, 134, 163, 232
blood glucose 250
blood osmolality 84
blood pressure 232
blood volume 232
blood-brain barrier (BBB) 183
body core temperature 130
body fluids 135
body size 130, 211
body temperature 6, 7, 19, 26, 40, 54, 63, 69, 70, 71, 80, 82, 86, 87, 88, 94, 95, 110, 115, 118, 134, 135, 147, 152, 153, 154, 155, 156, 162, 164, 194, 204, 208, 210, 216, 217, 222, 230, 232, 235, 238, 254
body temperature in humans 103
brain dysfunction 236
brain stem 167
brain stem reticular formation 94
Brodie 5
brown adipose tissue (BAT) 64, 65, 66, 67, 68, 69, 71, 73, 75, 148, 164, 210, 213, 248
burrowing 224
bursting 96, 97
bursts 51

C fibers 183
cafeteria diet 215

calcium 135
camels 216
Cannon 5
cardiac output 197, 232
cardiac reserve 195
cardiovascular disease (CVD) 253
cardiovascular reserve 195, 197, 203
carotid rete 88
cellular changes 200
cellular fatigue 237
cellular mechanisms 199
cenothermy, also normothermy 18, 184
central neurons 94
central venous pressure 88, 232
cerebral edema 236
Chargas disease 75
chemical thermometers 274
children 138, 139, 239
chronic hypothermia 246
chronological age 162, 164
circadian 156, 217
circadian activity 102
circadian clock 102
circadian rhythms 103
circulatory failure 235
circumventricular organs 183
classical (CHS) 231
Claude Bernard 5
clinical hyperthermia 268
closed control loop 26, 34
closed-loop negative feedback system 19
clothing 86, 210
cold 51, 63, 69, 70, 72, 75, 87, 114, 116, 117, 119, 120, 121, 139, 147, 149, 152, 156, 162, 164, 165, 167, 208, 209, 211, 212, 223, 246, 253, 254, 263, 264
cold acclimation 74
cold adaptation 208
cold diuresis 252
cold injuries 246
cold receptors 94, 95, 96, 97
cold sensors 55, 218

cold stimuli 49
cold tolerance 69
cold-acclimated 64
cold-acclimatized 119
cold-adapted animals 49
cold-exposed individuals 73
cold-induced vasodilatation (CIVD) 117
cold-induced vasodilation (CIVD) 87, 252
cold-sensitive neurons 97, 100, 101, 102, 181
coma 167
comfort 111, 113, 114
comfort zone 40
comparative aspects 203
competition for shared effectors 260
composite core temperature 218
composite skin temperature 218
conductance 29, 211, 212, 213
conduction 26, 27, 86
conductive heat exchange 30
controlled cooling 217
controller 19
controlling system 38
convection 26, 28, 30, 131
convective heat exchange 86
convective heat loss 40, 53
convective heat transfer 28
core 14
core temperature 18, 49, 67, 84, 111, 112, 113, 116, 119, 129, 133, 136, 149, 150, 151, 163, 246
cortex 94
countercurrent blood flow 212
countercurrent heat exchange 28, 88
cryoanalgesia 268
cryosurgery 268
Currie 7, 8, 9, 10
cutaneous blood flow 80
cutaneous thermoreceptors 94, 95, 97
cutaneous vasoconstriction 101
cutaneous vasodilation 84, 88
cyclic-AMP 67

cytokines 155, 179, 180, 181, 183, 184, 261

deacclimation 202
defervescence 186
dehydration 102, 134, 135, 186, 197, 235
density of sweat glands 82
depolarizing prepotential 99
diabetes mellitus 167
diabetics 239
diazepam 165
diet-induced non-shivering thermogenesis 63, 75
diet-induced thermogenesis 215
disease 265
Djungarian hamster 211
dormancy 216, 217
drinking 102, 135
drinking behavior 84
drugs 168, 239, 266
dry 80
dry heat exchange 30
dry heat transfer 85
dry transfer coefficient 31
dynamic firing rate 95, 97

eccrine sweat (atrichial) glands 82
eccrine sweat glands 83
ectothermic 117
ectotherms 15
effective insulation 28
efficiency 128
elderly 119, 120, 139, 162, 163, 164, 165, 167, 168, 169, 187, 239, 254
electromyographic recordings 49
electronic thermometers 273
elephant 213
Ellis 5
emotion 116
endogenous antipyretic 155
endogenous pyrogen 178, 179, 181, 183, 184

endothelial cells   183
endothermic homeotherms   63, 149
endotherms   15, 208, 210, 211, 213, 215, 216, 224
endotoxin   236, 239, 240
endurance   194, 195, 203
endurance training   130, 136
energy metabolism   203
energy substrates   137
environmental stressors   265
environmental temperature   129, 131
error signal   113
erythrodermia   119
Eskimos   210
esophageal temperature   129, 276
estradiol   262
estrogen   102
Etruscan shrew   213
evaporation   27, 32, 42, 80
evaporation transfer coefficient   32
evaporative cooling   194
evaporative heat   36
evaporative heat loss   32, 41, 80, 82
exercise   114, 115, 121, 128, 129, 130, 131, 132, 133, 134, 135, 136, 138, 139, 162, 198, 199, 203, 213, 238, 240
exertional (EHS)   231
exogenous   184
exogenous pyrogens   178, 179, 181
extremity cooling   247

Fahrenheit   8
fastigium   186
fat   133, 134
feeding   102
fetus   155, 156, 157
fever   7, 18, 116, 121, 155, 165, 178, 180, 181, 183, 184, 186, 187, 188, 220, 222, 236, 265
fever lysis   186
forced convection   86
forced convective heat exchange   30

Franklin   5
free convection   30
free fatty acids   137
freezing cold injury (FCI)   251
frostbite   251

Galen   8, 10
Galileo   8
glucose   102, 137
gram-negative bacteria   179

habituation   116
hands   117
head-out thermoneutral water immersion   136
heart   195, 239
heart rate   130, 195, 196, 201
heat   53, 82, 87, 133, 156, 157, 163, 164, 169, 194, 195, 196, 197, 201, 203
heat acclimation   194, 195, 196, 197, 198, 200, 201, 202, 203, 204, 240
heat acclimatization   132
heat balance   29, 35
heat balance equation   33, 35, 40
heat content of the body   15
heat cramps   230
heat edema   169, 230
heat exchange   17, 25, 26, 34, 80, 85, 88, 131
heat exhaustion   231
heat gain   114
heat illness   162, 164, 167, 168, 230
heat loss   26, 80, 86, 88, 101, 114, 129, 131, 133, 139, 151, 154, 181, 186, 209, 210, 211, 232, 263
heat production   10, 26, 36, 48, 49, 53, 54, 63, 64, 66, 68, 100, 101, 114, 129, 131, 133, 134, 137, 148, 149, 150, 153, 162, 164, 181, 186, 208, 210, 213
heat shock proteins (HSP)   200, 241
heat shock response   241
heat storage   33

Index

heat storage rate 26
heat strain 230
heat stroke 122, 231, 232, 235, 237, 238
heat syncope 169, 230
heat-exchanging surface area 17
heat-related illnesses 135
helium 86, 265
heterothermy 215
hibernation 63, 211, 215, 216, 217
hibernoma 75
hidromeiosis 84
Hippocrates 7, 8, 9
homeotherms 15, 19, 48, 204, 230
homeothermy 35, 48
host defense 178
hot flushes 121
human 108, 136, 147, 150, 154, 164, 200, 210, 247, 250
human infant 72, 73
human neonates 151, 153
humans 88, 94, 109, 111, 116, 133, 135, 137, 165, 180, 187, 195, 201, 202, 203, 209, 212, 213, 222, 224, 236, 238, 240, 246, 264
hyperhidrosis 119
hyperosmolality 88
hyperthemia 18, 109, 110, 113, 114, 117, 120, 122, 151, 154, 178, 230, 232, 237, 238, 239
hypobaria 263
hypohydration 238, 263
hypothalamic disorders 166
hypothalamic temperature 54
hypothalamic tissue slices 102
hypothalamus 55, 88, 95, 208, 213, 218, 220
hypothermia 18, 110, 115, 117, 122, 148, 149, 151, 154, 162, 163, 164, 165, 166, 167, 168, 215, 216, 246, 247, 250, 251
hypothermic 112
hypothyroidism 166
hypovolemia 88

hypoxia 263, 264, 265

iatrogeny 117
ice-cold water 117
ideal radiator 31
IL-1 262
IL-1 receptor antagonist 180
IL-1β 179, 180, 183, 261
IL-6 179, 180, 183, 261
immature neonates 72
immersion hypothermia 246
immobility 167
infantile febrile convulsions 188
infection 155, 187, 188
infectious agents 178
insensible perspiration 82
insidious (silent) hypothermia 254
insulation 133, 149, 154, 168, 208, 209, 210, 211, 212, 223
interferon 179
interleukin-1 ra 155
interleukins 179
intermediolateral column degeneration 167
internal body 14
interthreshold zone 187, 216, 217
intracranial pressure 236
ITZ 220, 222, 223

James Currie 3
John Hunter 4

Kupffer cells 183

lactacidosis 235
Langley 5
latent 184
latent heat of vaporization 32
Lavoisier 4
lean 134
leprosy 118
Lewis reaction 87
lipopolysaccharide 179, 236

liquid-in-glass thermometers 273
load error 38, 39, 40
Lovat Evans 5
lower critical temperature 131, 133, 198, 246
LPS 180, 183

Magendie 5
malignant hyperthermia 117
malnutrition 168, 187
Malpighi 5
meal 114
mean body temperature 218
mean skin temperature 84
men 120, 121
menstrual 134
menstrual cycle 102, 138, 262
mental functions 251
mental work 116
metabolic acidosis 235, 251
metabolic aspects 52
metabolism 26, 35, 49, 53, 72
micro-climates 109
microglial cells 183
migration 108, 224
modulatory inputs 220, 222
motivation 111, 114, 115, 118, 119
muscular tremor 49
muscular work 115

NA 222, 223
NaCl 84
Na-K pump 97
negative feedback 38
neonate 64, 71, 109, 147, 149, 150, 153, 154, 155, 187, 210, 264
nerve conduction velocity 251
nervous control of shivering 53
nervous system 95
nest-building 224
net conductive heat flow 27
neuroleptic drugs 117

neurological disorders 166
newborn 49, 209, 110, 248
Newton's law of cooling 15
nitric oxide 200
non-evaporative 85
non-evaporative heat loss 80
non-freezing cold injury (NFCI) 251
non-shivering 63, 70, 133, 248
non-shivering thermogenesis (NST) 49, 63, 64, 65, 69, 74, 101, 149, 157, 210, 213, 264
nonshivering thermogenic capacity 73
noradrenaline 87, 212, 215
noradrenergic (NA) 220
norepinephrine (NE) 56, 64, 65, 67, 70, 71, 72, 74, 165, 183, 264
NST 223
nychthemeral 18, 122
nycthemeral cycle 113

O_2 consumption 198
obesity 75, 122, 162, 168
obligatory non-shivering thermogenesis 63
obligatory thermogenesis 199
occupational hypothermia 247
ontogeny 155
operant behavior 108, 109
operant conditioning 110
operative temperature 31
oral contraceptives 138
oral temperature 276
organum vasculosum laminae terminalis (OVLT) 183
osmolar concentrations 135
osmotic diuresis 250
other internal body temperatures 277
other thermometers 274
ovarian cycle 113, 121
overall heat transfer 29
ovulation 102
ovulatory cycle 262
oxygen consumption 64, 129

pain   118
panting   32, 101, 135, 204
paralysing diseases   167
paraplegics   118
parasympathetic innervation   66
Parkinson's disease   167
passive system   34, 35, 36, 40
pathophysiology   184
patterns   187
perception   112, 167
perinatal period   65
PGE_2 183, 184, 262
phaeochromocytoma   75
phase of falling Tc   186
phase of rising Tc   184
physical fitness   136
physiological age   164
piloerection   63, 86, 212
plasma osmolality   102
plasma protein mass   198
plasma sodium   135
plasma volume   197, 198, 201, 235, 238, 240, 251
plateau phase   186
pleasure   113, 114
PO/AH   97, 102, 103
PO/AH cold-sensitive neurons   99
PO/AH thermosensitive neurons   100
PO/AH warm-sensitive neurons   99, 101
POA   181, 183, 184, 261, 262
poikilotherms   15, 204
postural adaptation   108
postural changes   223
potassium A current   99
precocial neonates   71
precooling   134, 139
pre-exercise cooling   115
pregnancy   187, 239
premature infants   153
preoptic and anterior hypothalamus (PO/AH)   95

preoptic area of the anterior hypothalamus (PO/AH)   6
preoptic area of the hypothalamus   200
preoptic-anterior   55
prescriptive zone   131, 134
preterm infant   154
primary motor center   55
prodromal period   184
progesterone   102, 262
prolonged fevers   188
prolonged hypothermia   246
proportional control   39
prostaglandins   222
psychotic   118
pyrexia   187
pyrogens   100, 155, 178

quinine   9

radiation   27, 31, 85, 131
radiation thermometers   274
radiative heat loss   41
radiative heat transfer   31
Ranson   5
rectal   115
rectal temperature   129, 275
relative humidity   41, 42, 263
reproductive hormones   102
resistance   29, 208, 210
respiratory evaporative heat loss   32
respiratory heat loss   40
respiratory tract   82
resting metabolic rate   264
reticular formation   95
rewarming   250
Robert E. Smith   64
rodents   64

salicylates   9, 117
saliva spreading   82

salivation   195, 203
Sanctorius   8
schizophrenia   118
secondary controlling centers   55
selective brain cooling   88, 120
sensation   111, 113, 118, 186, 209, 252
sensitivity   101
sensory perception   164
sepsis   187
serotonergic (5-HT) neurons   220
set-point   6, 19, 64, 101, 110, 111, 113,
    116, 117, 120, 122, 184, 186, 219, 220,
    252
shivering   16, 36, 48, 49, 49, 51, 52, 53, 54,
    55, 63, 69, 73, 74, 101, 111, 113, 115,
    133, 134, 150, 164, 210, 213, 216, 217,
    222, 223, 248, 252, 264
shivering ("chills")   186
Shy-Drager syndrome   167
signal transduction pathways   199
skin   82, 94, 95, 100, 101, 111, 119,
    131, 132
skin BF   235, 238, 240
skin blood flow   85, 87, 89, 131, 136, 139,
    167, 194, 195, 212, 264
skin blood vessels   88
skin circulation   130
skin perfusion   240
skin temperature   16, 41, 54, 80, 85, 87, 94,
    101, 110, 112, 113, 116, 117, 120, 131,
    133, 139, 149, 151, 278
skin thermoreceptors   164
skin vasodilatation   116
skin vasomotion   117
sleep   114, 115, 187, 216
sleep-wakefulness cycle   261
social behavior   109
solar radiation   263
somatosensory   94
spherocytes   237
spinal cord   55, 100, 101, 167
splanchnic   195, 232

splanchnic BF   235
splanchnic blood flow   240
splanchnic perfusion   200, 235
sprint running   140
stable   186
static firing rate   95, 97, 96
Stefan-Boltzmann radiation constant   31
Stone   9
strategies   211, 215, 224
stress   111, 187
stroke volume   130, 195, 197, 201
strokes   162, 166
suprachiasmatic nucleus   102
surface area-to-mass ratio   147, 154
sweat   41, 82
sweat composition   84
sweat glands   84, 119, 120, 151, 164, 200
sweat rates   82
sweat secretion   32, 82
sweating   6, 88, 101, 111, 115, 116, 130,
    131, 132, 134, 135, 138, 139, 151, 152,
    154, 164, 167, 186, 195, 203, 212, 232,
    239, 262, 263
sweating rate   136
swimming   133
Sydenham   8
sympathetic   64
sympathetic activaton   72
sympathetic cholinergic stimulation   84
sympathetic hyperactivity   75
sympathetic innervation   66
sympathetic nervous system   55, 64, 65, 87

tachymetabolism   15
temperature   14, 28, 39
temperature measurement   18
temperature safety margin   240
temperature-insensitive neurons   99, 100
test drops   217
thalamus   94
themoregulatory nonshivering
thermogenesis   63

therapeutic hyperthermia and hypothermia   268
thermal boundary layer   263
thermal comfort   111, 112, 116, 119
thermal conductivity   30, 44, 86
thermal core   16
thermal gradient   17
thermal homeostasis   19
thermal insulation   29
thermal micro-climates   108
thermal muscle tone   51
thermal neutral environment   94
thermal preference   121
thermal sensitivity   217
thermal shell   16
thermoeffectors   19
thermogenesis   70, 119, 133, 213, 248
thermogenic activity   70
thermogenin   68, 215
thermometer   7, 8, 273
thermometeric measurements   274
thermometric device   18
thermoneutral   17, 39, 64, 72, 82, 85, 209
thermoneutral conditions   131
thermoneutral environment   70, 187
thermoneutral zone   131, 154, 198, 209
thermoneutrality   18, 263, 264
thermoreceptors   34, 38, 94, 98, 100, 149
thermoregulatory nonshivering thermogenesis   75
thermosensitive neurones   6, 102, 261
thermosensitivity   97, 98, 101, 102
thermosensors   19, 208, 209
thermotolerance   240
third ventricle   95
threshold   219, 220, 240, 262
thyroid   198, 199, 213
thyroid gland   119
tissue conductance   36
TNF-α 179, 180, 183, 261
tolerance   114, 133, 194, 200, 208, 216, 236, 238, 239, 240

torpor   63, 215, 216
trench or immersion foot   252
tumor necrosis factor   179
tympanic   115
tympanic temperature   277

UCP   69, 70, 71, 72, 73, 74, 75, 76
uncoupling protein (UCP)   68, 149
upper critical temperature   80, 131, 132, 133

vagal afferent nerves   183
van't Hoff-Arrhenius law   15
vapour pressure   42
vascular tone   87
vasoconstriction   63, 133, 163, 186, 209, 212, 248, 264
vasoconstrictor activity   87
vasodilatation   163, 194, 201
vasodilation   186
vasomotor   115
vasomotor changes   36
vasomotor response   119, 152
vasomotor tone   6
VO_{2max} 130, 136, 198
volumetric heat flow   28
voluntary dehydration   238
warm   114, 120, 132
warm acclimation   115
warm receptors   94, 95, 97
warm sensors   218
warm-acclimatized   120
warming-up   134, 139
warm-sensitive   181
warm-sensitive neuron   97, 98, 100, 102
water   86, 133, 134
water retention   102
water vapor pressure   41, 263
weighting factors   218
Wernicke's encephalopathy   167
wetted area   32, 42
white adipose tissue   65

white fat cells   73
whole body cooling   247
wind   263
wind chill   247
winter mortality   253, 254
women   134, 138, 102, 120, 121, 262
work load   129
Wunderlich   8, 10

yoga   116